HORNED BEETLES

HORNED BEETLES
A Study of the Fantastic in Nature

by the late
GILBERT J. ARROW

edited by
W. D. HINCKS
Manchester

Dr. W. JUNK, PUBLISHERS — THE HAGUE — 1951

Copyright 1951 by Uitgeverij Dr. W. Junk, Den Haag.

Printed in the Netherlands.
Drukkerij Gebr. Verweij, Wageningen.

FOREWORD

GILBERT JOHN ARROW was born near London in 1873 and in his 23rd. year joined the staff of the British Museum (Natural History) as an entomologist.

The whole of his long life was devoted to the study of beetles and particularly to the great group of Lamellicornia to which most of the 'Horned Beetles' referred to in this, his last work, belong. As long ago as 1899 ARROW wrote on the remarkable sexual dimorphism of some members of this group, a subject to which he constantly returned during his extensive studies of the extraordinary insects which had so captured his interest. The reason and purpose behind the amazing development of 'horns' in many male Lamellicorns and other beetles, and their extreme variability has long baffled biologists, including the great DARWIN himself! Thus the mature thoughts on this subject of a man who devoted the whole of a long life to its study are of special significance and are well summarized in the present little book on 'Horned Beetles'.

Not only does this work provide an unusual topic for the general reader and a challenge to the field naturalist but it is important to the serious student of coleopterology. To make it more acceptable to the last, without detracting from its appeal to others, a scientific index has been added, indicating the authorities, synonymy etc. of the scientific names appearing in the text. To have added this information as footnotes would have been burdensome for the layman. An authors' index, subject index and selected bibliography have also been added, but otherwise, with the exception of a few minor alterations, the text of 'Horned Beetles' has been left as it was at the death of its author, on October 5th. 1948.

Many younger coleopterists have cause to remember the kindness, practical help and encouragement which ARROW gave them, amongst their number the present writer, who has before him as he writes a thick file of ARROW'S letters testifying to his crisp humour, sound advice and ready help over a period of many years. In saying 'Vale' to such an old friend it is a privilege to be able to do something, even so little, to introduce his last work on 'Horned Beetles'.

W. D. HINCKS,

Manchester Museum, Nov. 17th. 1950

The Editor wishes to thank Mr. E. B. BRITTON of the British Museum (Natural History) for the photograph appearing on Plate 3.

CONTENTS

CONTENTS

PREFACE

This little book is the outcome of many years' study of its subject prosecuted chiefly in the Natural History Museum, South Kensington. It owes much to the suggestions and criticisms of various friends and colleagues. Amongst these I would especially mention SIR GUY MARSHALL, F.R.S., late Director of the Commonwealth Institute of Entomology, MR R. I. POCOCK, F.R.S., late Superintendent of the Zoological Gardens, London, DR A. T. HOPWOOD, DR K. G. BLAIR, DR F. ZEUNER, Dr F. VAN EMDEN, colleagues in the Natural History Museum, all of whom have been most helpful.

As to the photographic illustrations, it should perhaps be mentioned that none are taken from living specimens. Although some are represented in lifelike attitudes, these have been selected only to show their form to the best advantage and should not be regarded as affording evidence of their conditions of life.

1

CHAPTER 1.

INTRODUCTION

"If we could imagine the male *Chalcosoma,* with its polished bronze coat of mail and its vast horns, magnified to the size of a horse or even of a dog, it would be one of the most imposing animals in the world." This was written of a beetle by DARWIN in his book "The Descent of Man" and all who see either that or another one of the great horned beetles for the first time cannot fail to experience feelings of astonishment. The size of these giant insects ensures that their fantastic shapes shall not be overlooked but great numbers of smaller kinds wholly unknown to the world at large exhibit forms as strange or even stranger and a general survey of the whole field, impossible in earlier days, reveals an astonishing variety of curious forms such as DARWIN himself could not have suspected. The question how and why such extraordinary forms have been acquired is constantly asked but never satisfactorily answered. The explanation generally accepted, that the "vast horns" are weapons, will not account for their endlessly varied and often grotesque forms or for their presence in some and absence in other beetles essentially similar in other respects. Reflection upon the problem leads to the realization that the same difficulty exists with regard to the horns of other animals, which show the same curious characteristic — a tendency to exhibit a grotesqueness and extravagance incompatible with the state of efficiency required of weapons employed for any important purpose. The discovery that these organs are usually either peculiar to the male or more highly developed in that sex at once relates them to many no less puzzling male features, such as the horny casques borne by some of the strangest of birds, the hornbills, etc., the fleshy combs and wattles of others, the crests of some lizards, newts and fishes, and many other strange accompaniments of the male sex, including the often gorgeous and sometimes fantastic plumage of so many male birds. The problem with which the horned beetles confront us is in fact far-reaching and concerns a very large part of the animal kingdom and many of its most extraordinary forms.

When any form of life appears to human eyes grotesque or fantastic this is generally due to the proportions between its different parts not being such as we have learnt to regard as customary. It may be the result of adaptations to conditions of life unimaginably different from those with which we are familiar, as in some of the monstrous reptiles of the remote past or the fishes existing to-day in the profound unlighted depths of the ocean. But the strange forms with which we are concerned cannot be explained in any such way, for the conditions in which they live differ in no important respect from those of others normally proportioned, and some quite different explanation must therefore be looked for. It was to this problem that DARWIN devoted the greater part of his book "The Descent of Man" and the result of his study of the problem was his theory of sexual selection. The evidence bearing upon the subject nearly a century ago was far less extensive than that available now and those who are familiar with even a small part of the facts since discovered upon the subject of sexual dimorphism (the differences of aspect between male and female animals) have realised that the theory of sexual selection as put forward by DARWIN is inadequate for their explanation.

Although beetles only provide one chapter of the evidence in a very wide field of enquiry that chapter is an important one, the number and variety of the examples provided being far greater than can be found in mammals, birds, reptiles, fishes, crustaceans or any other group and this great body of evidence is as yet almost unexplored.

About a quarter of a million beetles of different species have been named and described and it cannot be doubted that many more than that number remain undiscovered for, so far as insects are concerned, the world is yet by no means half explored. Of this great multitude the horn-bearers amount to very many thousands and provide a copious reservoir of facts for the study of a problem which has not ceased to exercise many minds from the time of DARWIN to the present day.

What is a horn? Originally, no doubt, applied to the hollow outgrowths borne by cattle, the term has by common consent been extended

3

to cover a very great variety of appendages of a more or less similar character and it will be well to realize that it is not a scientific designation capable of any rigid limitation. If we restrict the term to such organs as are composed of the substance specifically called horn we must exclude those of many kinds of animals, including all the insects, to which in common speech it is applied. Even among mammals horns differ widely in their nature and origin. We cannot even confine the name to protuberances borne upon the head for in insects structures identical in character are found upon other parts of the body. In beetles some of the most familiar examples of horns so-called are in fact the jaws.

When, however, we consider the very various structures which ordinary usage entitles us to group together as "horns" we shall find that they possess a number of features in common and can, without any confusion of ideas, be treated as a whole. Their appearance is generally an indication of approach to maturity and, although they may be possessed by both males and females, they are almost invariably either only developed or more developed in the former. Their growth is usually rapid and they have a noticeable tendency to attain a size which, in relation to that of their bearers, can only be described as extravagant. These characteristics they share with other distinctively male features collectively known as secondary sexual characters and it is therefore impossible to discuss horns without referring to others of these. The peculiarities found in the organs ordinarily called horns are also found in others, such as the tusk of the small whale, the Narwhal, which, placed by some imaginative artist upon the head of a horse, became the horn of the unicorn in the Royal Arms of England, as well as the tusks which Nature has placed upon the top of the head of that strange pig, the Babyrusa. The tusks of the elephant, as well as the jaws of the stag-beetle, are fundamentally insignia of the adult male and subject to the same laws of growth.

Before proceeding to deal with horned beetles in particular, it will be useful to describe the characteristics of beetles in general, constituting the Order Coleoptera, in order that the meaning of such technical terms as it is necessary to use may be clearly understood; and to give a brief survey of the more important groups of insects included in that order to show roughly the distribution amongst them of horned forms.

4

Beetles, unlike locusts and grasshoppers, cockroaches, bugs and various other insects, do not develop by uniform stages from the moment of leaving the egg to the mature condition, but, like butterflies and moths, bees, ants, flies, etc., leave the egg as grubs entirely unlike their parents and, after feeding for a period which varies from a few weeks to a few years, attain their full growth and only after undergoing an inactive chrysalis (or pupa) stage, reach their final adult condition. The grub, or larva, is in some cases, as in the ladybirds, an active six-legged little being, capable of hunting for its own food, and in other cases an inactive, sometimes legless and often blind creature, with only slight power of movement, obtaining food only by virtue of the parental instinct which has ensured that it shall begin life in proximity to a sufficient supply of the particular kind of food it needs. In certain exceptional cases, as in the Melöid beetles, which are parasitic upon bees and wasps, the newly born insect is very active but, after attaching itself to a wasp or bee and, if fortunate, being carried to its nest, it there becomes transformed into a helpless maggot and in that form devours the store of honey, pollen on other food provided by the victimized insect for its own offspring. Others of these Melöids devour the underground egg-clusters of locusts.

The mature beetle, like other insects, is a six-legged animal with a segmented body divided into three regions, — the head, which carries the sense-organs, eyes and feelers (antennae), as well as the various organs of the mouth, including two pairs of jaws, the upper pair known as the mandibles and the lower pair called the maxillae; the thorax, bearing the organs of locomotion, legs and wings; and the abdomen, containing the principal viscera. From the other orders of insects beetles differ primarily in the fact that the first of the two pairs of wings, which they possess in common with most other insects, have ceased to serve the purpose of locomotion and become hardened into wing-covers (elytra) to protect the thin membranous second pair. The latter, when not in use, are folded up beneath the elytra. It is owing to this admirable arrangement that some beetles are able to spend most of their lives in water, others to burrow in the ground or into solid wood and other substances without injury to their delicate flight-organs and that this Order of insects has been

5

able to adapt itself to almost innumerable modes of life, becoming so exceedingly prosperous and abundant.

The legs of beetles, like those of other animals, are composed of several joints, the principal parts being the thigh (femur), the shank (tibia) and the foot (tarsus), the last divided into from three to five small joints and bearing a pair of claws at the end.

It is important to realize that in insects all growth takes place in the preliminary stage of development and that, once maturity is attained, the rigid skeleton, which is external, not internal, is an absolute bar to any further change. The size attained on maturity may vary enormously, so that a series of beetles, or other adult insects, although all belonging to the same species and even to the same brood, may appear to represent many different stages of growth, but, since adult insects cannot grow, the size of each, whether large or small, is fixed and final, if we except the more or less flexible abdomen which, in certain cases, is subject to considerable distension.

All the appendages of the perfect insect, legs, wings, antennae, as well as the horns, however fantastic in size and shape these may be destined to become, are formed within the skin of the grub when it has reached its full size and are seen, when the skin is shed, compactly folded together in the quiescent pupa. At the end of the resting stage a final moulting of the skin which envelopes every part starts the astonishing process, like the unfolding of a flower, by which, from the pale, shapeless, mummy-like object, the vivid, armour-clad, active beetle gradually takes shape and colour. This wonderful process has never yet been recorded in any horned beetle. In such magnificent creatures as the great *Chalcosoma* (Plate 1), or the still greater Hercules-beetle (Plate 2), or the huge-jawed *Macrodontia cervicornis* (Plate 3), it must be a truly remarkable sight.

Since the hardness of the exterior precludes all further growth in the adult insect, once this stage is reached the taking of nourishment becomes in most of them comparatively unimportant. The life of many adult insects is very short, some are even without organs for taking food, others, provided they are not exposed to evaporation, live for months without needing any and nearly all feed on a much reduced scale compared with that of their period of growth.

6

The substances which serve as food for the multitudinous different kinds of beetles are of almost infinite variety. Very many of them feed upon carrion or decaying matter of all sorts, often after burying it, and do useful service by removing it from the surface of the ground. Particularly useful are the many different kinds that attack fallen trees and tree-stumps in forest regions, hastening their decay, and so make way for fresh vegetation.

So important are many of those beetles which prey upon other and harmful kinds of insects that their transfer from one country to another by artificial means has become a regular and highly effective method of dealing with the calamitous attacks of pests accidentally imported into regions where their natural enemies are absent.

A few words must be said in explanation of insect names. Since it is now recognized that the number of existing kinds of insects must amount to millions, it is evident that the classical and modern languages together would not suffice to provide a different single name for each. Each species is therefore given two names. The first (the generic name) it shares with the others to which it is most nearly related; the second (the specific name) is its own, although it may occur again in other genera. A number of genera forming a natural group are called a family and given a distinctive name, which always ends in -idae (e.g., Carabidae, Lucanidae); while the termination -inae is used for subdivisions of the family (Cetoniinae). Horned beetles do not form a natural group, but are found in larger or smaller numbers in many different groups, all of which contain hornless forms as well. In many large groups no horned forms occur and in others they are very rare. It is only in a few families that they are numerous.

A rapid survey of the most important groups of beetles will serve to show roughly the relation borne by the armed minority to the unarmed majority. There are the fierce Tiger-beetles (Cicindelidae), swift-running and swift-flying, which, in their adult stage, hunt and devour other insects and, in the larval stage, lie in wait at the mouth of their tunnels to seize unsuspecting passers-by. There are the predacious Ground-beetles (Carabidae), also swift-footed and voracious, hiding under stones and rubbish by day and hunting chiefly at night, ubiquitous throughout the world and multitudinous in their species. Various other

7

families of beetles are predacious to a greater or lesser extent, but the Carabidae are the Carnivora par excellence of the insect-world. Many of the lithe, slender Staphylinidae prey upon other insects and many of the short, smooth, slow-moving Histeridae seek out in their burrows and devour helpless grubs unable to escape them. The Glow-worms and Fireflies (Lampyridae) are the special foes of snails and the related Cleridae prey upon the grubs of wasps and bees etc. In none of these groups do we meet with horned species, with a very few rare exceptions, such as the burrowing Staphylinids, *Bledius,* some of the males of which bear upon the head and thorax slender horns directed upwards or forwards and no thicker than fine flowerwire. There are no horned forms amongst the Carnivorous Water-beetles (Dytiscidae) nor amongst the Vegetarian Water-beetles (Hydrophilidae), an unrelated although superficially similar family.

The immense assemblage of the Heteromera, which includes the Meloïdae, already mentioned as preying parasitically upon the larvae of bees and wasps, contains a number of horned forms (*Phrenapates, Atasthalus,* Plate 4, fig. 14), which all appear to subsist upon rotten wood or the woody fungi that grow upon trees. The giant amongst all the Heteromera is the great Himalayan *Autocrates aeneus.* The males of this, like those of the giants of various other groups, have their mandibles grotesquely enlarged.

It is when we reach the group of families constituting the suborder Lamellicornia (the chafers or leaf-horned beetles — the term horn is here misapplied to the antennae) that we find horned insects in large numbers. Many of these, like the Stag-beetles (Lucanidae) feed, in their immature condition, upon decaying wood, in dead trees or stumps, and, when adult, upon exuding sap or other plant-juices. Others, such as the gigantic Dynastinae (Rhinoceros beetles, etc.) and Goliath beetles (Cetoniinae), find their nourishment in roots or vegetable debris and a very large number (Geotrupinae, Coprinae) are dung-feeders like the Dor-beetles (*Geotrupes*) and related kinds, whose nest-building and food-storing habits have been described in the writings of J. H. FABRE.

The remaining great groups of Coleoptera are entirely vegetarian. They include the Cerambycidae, Longicorns or Long-horns (referring again to the antennae), timber-destroying insects, many of the largest of which have the jaws of the males exaggerated in the same way as those of the Stag-beetles; the Curculionidae or Weevils, which feed upon the leaves, fruits, stems or other parts of plants of all kinds, and a few of which have very curious horns in the male, springing, not as usual from the upper, but from the lower surface of the body (*Mecopus,* etc.) (Plate 4, fig. 7); the very slender, stick-like Brenthidae (Plate 4, figs. 1—6), wood-borers like the Cerambycidae, of which many of the males, as in that family, have enlarged mandibles; and finally the Chrysomelidae, leaf-eaters nearly all, a vast host abounding in curious patterns and gorgeous colours. Most of them are of small size and none can be called hornbearers, although some of larger size than most (e.g., in the subfamily Clytrinae) show an enlargement of the head and mandibles of the males which, if still further exaggerated, would raise them to that rank.

From this very cursory survey it will be noticed, first, that, as in animals of a higher grade, horns in beetles are generally, although not always, a mark of the male sex; and, secondly, that the horned Coleoptera are eminently pacific in their general habits, the predacious families are conspicuously absent and the great majority are either scavengers by occupation or feed upon timber, alive or dead. Their horns, formidable as they often appear, and sometimes, especially in those cases in which they are the enlarged jaws, bear a close resemblance to the gripping jaws of predacious insects of other Orders, such as the Ant-lions, are in no known case used for seizing prey. A very slight examination of many of their forms is sufficient to show that they would be quite useless as weapons. Many theories have been put forward as to their real significance, but before considering these it will be necessary to consider more carefully the various types of structure that occur and the habits of the insects bearing them.

CHAPTER 2.

HORNS AND SIMILAR FEATURES IN BEETLES

The structures to which the term horn is generally applied in beetles form only one group, not very well-defined, in a multitude of peculiar and puzzling features, generally, but not always, taking the form of out-growths from some part of the insect's exterior, and on almost any part of it. The leading characteristic is their appearance in the male on approaching maturity. They may occur also in females but almost always in a lesser degree of development than in the male; more often they are absent or appear only in a rudimentary form. Cases in which similar structures are found in females but not in males are so rare as to empha-sise the overwhelmingly male character of the phenomenon in general. In all this there is an exact parallel with many equally strange and puzzling features to be found in other animals.

It is true that horns are possessed by the grubs of some beetles but these invariably disappear as soon as the grub-stage ends. They are to be found, for example, in the grubs of certain Ground-beetles (Carabidae) which seize and devour other insects; in these (the larvae, for instance, of *Nebria, Galerita,* and other genera) they seem to be accessories for retaining a hold upon their prey. They also occur in the larvae of some Water-beetles, also of predacious habits. But these larval horns are never retained when the grub reaches maturity and those acquired by the beetle on reaching maturity are never found in its grub. Like the horns of higher animals, the first appearance of such structures always indicates approa-ching maturity. When the growth has been completed and the last larval skin is about to be cast off, the insect, lying in its pupal cell, always assumes a particular attitude in which the transformation to the adult form, a highly critical process, may be most safely undergone. Even the hardest parts of the future mature insect are at this time in an extremely soft state, in which the slightest pressure may produce malformation, and in many insects special spines or bristles upon the pupa prevent contact at any point with the cell-wall. It was first noticed by the celebrated French naturalist, J. H. FABRE, that in two French species of the great genus *Onthophagus* which he found in this stage, a short horn appeared upon the middle of the thorax, although there was no trace of it in the adult insects. MR HUGH MAIN (*Proc. Ent. Soc. Lond.,* 1922, p. 3) has found that immediately before casting its skin the larva of one of these two forms, *Onthophagus taurus,* turns over so that when the skin parts the body is supported upon this horn and kept from any other contact with the cell-wall. It is very probable that this movement is common to all the hundreds of species of the genus and the pupal horn as well. FABRE was not aware that, although the adult beetles of those species in which he found the pupal horn were without any corresponding process in the same situation, many Onthophagi in other parts of the world, such as the Indian *O. imperator,* (Plate 5, fig. 12), have a long upstanding horn there. The pupae of these species have not been discovered and it may perhaps be found, when they are known, that this is not developed from the pupal horn but has a different origin. Since all the adult features of insects, however, always appear first in the pupal stage it is perhaps more probable that the horn upon the thorax of *O. imperator* corresponds to the pupal horn that has been found in *O. taurus* and other European species and will occur, we cannot doubt, in the pupae of other kinds of *Onthophagus*. It is very remarkable that this horn disappears in all the adults of *O. taurus* and in *O. imperator* persists in the males alone. Unless the pupa of *O. imperator* carries upon its thorax both an impermanent horn and a permanent one, which does not seem likely, we must suppose the pupal horn to indicate that in ancient times the ancestors of these beetles, both male and female, bore a horn upon the back. The theory has been put forward that horns were originally common to Lamellicorn beetles of both sexes and confirmation of this appears to be provided by certain exceptional cases in *Onthophagus* and other genera. For example in the very curious little Indian *O. sagittarius* (Plate 5, figs. 13 & 14) the female bears a strong horn upon the thorax and another upon the head but the strikingly dissimilar male has two horns upon his head and none upon the thorax, the thoracic horn having swollen into a great hump.

From what we know of the habits of these burrowing and nest-building

11

insects, whose tasks are performed either by the female alone or by a male and female together, such a horn as that of the male *O. imperator* must render it impossible for him to collaborate, but the suppression of the impediment in the female enables her to perform all the necessary operations. Why, in *O. sagittarius,* the female has retained, so exceptionally, this primitive feature we cannot tell but the male seems to have exchanged it for a more practical form, for it has become enlarged to a hump with a broad flat surface, convenient for pushing out the earth which must be removed from the burrow in the operation of nest-building. We shall find other evidences of changes of this kind which must have occurred in past ages.

But if in some beetles horns were first acquired by both sexes and afterwards retained by one only, it is scarcely possible to believe that has been the common case, so impressive is the mass of evidence for a special connection between maleness and horn-bearing. We cannot dissociate horns from other features of a similar kind and often outgrowths from various parts of the body are not called horns only because they are not borne upon its anterior part. In beetles such outgrowths are of almost endless variety. They may be common to both sexes but far more often they are peculiar to the male and to find them in females and not in males is extremely rare.

One of the rare cases is a little European beetle, *Valgus hemipterus* (Plate 4, figs 8 & 9), whose female carries at the end of the body a long slender spine which looks like the sting of a wasp or bee; but, unlike a sting or the non-retractile egg-tube of female ichneumon-flies and other insects, it is not a tube but a horny extension of the last dorsal segment of the body. Many beetles have a tail of a similar kind but, when a difference exists between male and female, it is the male which has the longer appendage, as the Cockchafer. But the male *Valgus* has no tail at all. It is probable that the instrument of the female is employed in placing the eggs in the rotting wood beneath the surface of the ground in which the larvae live and feed.

The females of certain kinds of those remarkable insects called Ambrosia-beetles bear a very strange apparatus upon their heads but the only British representative (*Platypus*) is not one of them. These beetles

12

have the remarkable habit of cultivating in their burrows tender fungi, fancifully called Ambrosia (the food of the gods), upon which their young are nourished. They live in communities and excavate the burrows in the interior of decaying trees or logs. The young larvae move freely about the tunnels in some cases and in others inhabit separate cells dug out by the beetles, which crop the cultivated food and supply it periodically to their offspring. The fungus appears to be grown upon specially prepared forcing-beds within the galleries but the manner in which it is introduced and grown has not yet been discovered. It is believed that the apparatus upon the heads of the females of some of the various kinds is concerned in the operations. The head in these females is deeply hollowed in front and from all sides dense fringes of stiff reddish-golden hairs directed towards the centre cover and screen the hollow. Within the receptacle formed in this way have been found bunches of dried filaments recognizable as composed of the minute fungus cultivated by the insects (Strohmeyer, *Entom. Blätter*, 7, 1911, p. 103). The protecting fringes are long and curling and may be likened to the fans of ostrich-feathers which were fashionable at one time. The hairs may spring directly from the edges of the cavity or from special projections. In species of the genus *Crossotarsus* the first joint of the antenna is long drawn out in both sexes but the process is exaggerated in the female and carries one of the golden fringes. In some species the top of the head on each side near the eye has a projecting brow, which is also fringed. The most remarkable of all is the Sumatran *Spathidicerus Thomsoni*, the female of which has a deep cavity fringed with scaly hairs behind each eye, the basal joint of the antenna gives rise to a very large, nearly round, feathered screen, while each mandible is prolonged by a horizontal extension, which is dilated in front and curves upward and is covered with scaly hairs. This prolongation of the jaws of the female makes it impossible that they can be employed for boring into solid wood as usual and accordingly it has been suggested that the excavation of the galleries must be performed by the males alone of this species. It is interesting to find that the female has a well-developed posterior scoop, like so many males, by which means they push out of the burrow the wood-fragments dug out by the female. We can hardly avoid the conclusion that, in this strange species, the parts

13

usually played by the two sexes have become reserved, the labour of excavation being performed entirely by the male, while the female undertakes the ordinary male duty of removing the debris.

So very exceptional is the possession of these accessory structures by female Ambrosia-beetles that for a good many years the females were believed to be, and indeed were described and pictured as, the males and *vice versa*. No doubt other external structures to some extent of a similar kind exist in female beetles but they are certainly rare. The variety of such outgrowths in males, on the other hand, is bewildering and, most strangely, they are usually without any apparent use. Another noticeable peculiarity is that such male outgrowths may, in forms closely related, appear in one species upon one part of the body and in another upon some other part. For example, in a genus of little Oriental beetles called *Saula* an enlargement of the front tibia is found in the males of one species, while in another species it is the hind tibia which is enlarged, in a third it is the last point of the antenna and in a fourth all the joints of the antenna are thickened, so that the usual delicate flexibility of these organs is lost. So erratic are these features in their occurrence that they may be found only in two or three members of a large group all the rest of which are destitute of anything of the kind. Amongst the very numerous related beetles belonging to the family Tenebrionidae which crawl upon the dry veldt of South Africa the males of one, *Calognathus Chevrolati,* has its jaws elongated like those of an Ant-lion, while in all the rest they are very small and alike in both sexes. In all the hundreds of species forming the fungus-haunting family Erotylidae two or three only, belonging to the genus *Cytorea,* are known to have special horny processes in the male.

The wonderful African Reindeer-beetle, *Onthophagus rangifer,* of which the male bears trailing antlers almost as long as the body, (Plate 5, fig. 9) has also in that sex a stout hooked process projecting upward from each shoulder, as if to guard the delicate horns which, in the resting position of the head, lie upon the back close to the shoulders. In the very nearly related *O. gibbiramus,* (fig. 7), which has equally slenders antlers, there are similar hooks, placed at the hinder angles of the thorax instead of the shoulders. Other antlered species of the same group have no hooks at all.

14

Very strange appendages are those borne by males of certain beautiful glossy green Rose-beetles, chiefly found in Borneo, belonging to the genus *Pseudochalcothea*. In one species each hind tibia has a long and very slender ribbon of horny substance trailing back from the base and tapering to a very fine point, another has a similar filament, which instead of being pointed is knobbed at the end, a third has a thin rod thickened at the end and a fourth a very sharp hook. In other species closely related these appendages are quite absent and females are always without them. The fragile filaments, arising just below the knee, seem always to be in perfect condition, although they appear so delicate that injury must result from use in any kind of way.

In many other beetles the legs of the males bear horny teeth, spines or embellishments in one part or another (see Plate 11, fig. 4 & Pl. 12, fig. 7). The splendid South American *Chrysophora chrysochlora,* a glittering golden-green insect, with fiery red and gold legs, has the hindmost pair elongated and the tibia armed with a great curving spur at the end. In some species of the long-legged spider-like *Sisyphus,* named so from their habit of trundling a ball of food-stuff along the ground like the related *Scarabaeus,* the tiny joint called the trochanter, at the base of the hind leg, is drawn out into a long knife-like blade. but in most of the species this is absent. All the legs may be much longer in the male than in the female, as in the extraordinary Stag-beetle, *Chiasognathus,* Plate 9, figs 1 & 2, or one pair, as in the gigantic *Euchirus longimanus;* or on the contrary the legs may be shorter and stouter. Often the antennae are the seat of the difference. An increase in the sensory surface of these important organs, which adds to their efficiency, is very usual in males but the development is as often in the non-sensory part. In *Chiasognathus* the first joint, which bears no part of the sensory surface, is about three times as long in the male as in the female, while the sensory terminal part scarcely differs, and in a weevil, *Mecomastyx,* Plate 4, figs. 12 & 13, found in the New Hebrides, the same joint may actually be fifteen times as long as in the female. Again this joint may be broadened instead of lengthened or, instead of the first, another one may be dilated.

The wing-covers may bear outgrowths in the male. In certain members of the curious family Brenthidae, whose extreme elongation makes them

15

appear at first sight like bits of stick, the two narrow wing-covers are drawn out at the hinder end into long slender, sometimes thread-like but quite rigid, processes, which may be much longer than the wing-covers themselves. This of course increases the un-beetle like appearance of the insects and might be supposed to be of value in helping to deceive their enemies. But why is such protection denied to the females, which are either without the processes (as in *Brenthus armiger,* Plate 4, figs. 3 & 4) or have them only in a rudimentary form (*Diurus forcipatus,* figs. 5 & 6)? The destruction of the female is more likely to endanger the succeeding generation than that of the male and it is in the female that special protective devices are most often found, as in the concealing colour of many hen birds whose males are conspicuous. The tail-processes of the male Brenthid, which, when the two elytra are separated for flight, diverge at a wide angle, must considerably hamper that operation, and it may be that they are suppressed in the female for that reason.

Males of the half yellow and half black *Nephrodopus enigma,* in addition to three upstanding erections upon the thorax, have a pair, sharp and thorn-like, upon the wing-covers, all together making a circle of horns upon the back; males of *Spathomeles decoratus* have a curious hook directed backwards upon the middle of each wing-cover. There are weevil species in which knobs or finger-like processes appear just before the tips of the elytra, sometimes with a tuft of stiff hairs at the end of each. In *Exorides equicaudatus* the two processes together carry a great bunch of long coarse filaments like a birch broom or a horse's tail.

Outgrowths peculiar to the male may be found upon the lower, instead of the upper, surface of the body. The Indian *Dorysthenes rostratus* has a sharp spike upon its chest, meeting the long tusk-like mandibles, and some Malayan weevils have curious horns arising in the same way. *Mecopus spinicollis,* Plate 4, fig. 7, with its long spidery legs, has a pair of horns also thread-like in their slenderness and *Acythopeus unicornis* has a single horn arising in the same way and curving upward. These are processes from the prosternum but others beneath the body may be outgrowths of the metasternum (*Pedanus*) or of the abdomen (*Ancistrosoma*). In *Adoretus celogaster* a row of little swellings upon the abdominal segments in a line

16

beneath the body are slightly prominent in the female but in the male become sharp conical prominences.

Many other strange modifications or outgrowths in other regions of the body and peculiar to one sex, almost invariably the male, might be enumerated but this incomplete list will be sufficient to show that there is no part in which, in one or another kind of beetle, such features may not occur. It will be noticed that in the few known instances of such structures in females there is good reason to conclude that a practical purpose is served by them; but it is quite otherwise in the case of males. In certain of these also definite uses can be discovered but, so far as we have the means of judging, most of these peculiar outgrowths, such as the knobs or hooks upon the wing-cases or their filamentous extremities, the leg-filaments of *Pseudochalcothea* and the head-filaments of the Reindeer-beetle, serve no purpose. In most cases it is observable that such peculiarities are confined to one or a few species and absent in nearly related forms, otherwise similar and of apparently the same habits. It may even be evident in certain cases that, although male beetles in general are more active and agile than females, these appendages have so far developed that, like the enormous horns borne by some of the goats, they have become actual encumbrances. This, strange as it may seem, is in accord with similar facts to be observed in the differences of colour and pattern often found between the two sexes. In very many beetles, as in many birds and other creatures, while the female enjoys what is known as protective coloration, that is to say, such as renders it difficult of detection by enemies; the male, on the contrary, is highly conspicuous. This is the case, for example, in the Atlas-beetle, *Chalcosoma.* The male is at once revealed, not only by his enormous horns and great size but by his glossy exterior and metallic green lustre, while the smaller female is very dull and difficult to detect. In innumerable other beetles the females are dull-coloured and the males brilliant or decorated with vividly coloured patches, stripes or patterns. Perhaps the most exquisitely coloured beetle in Europe, the silvery-blue *Hoplia coerulea* of Southern France and Switzerland, has a dull brown quite unattractive female. It cannot be doubted that, favoured as we may believe the males to be by their superior beauty, it is actually a danger to them and it is the females which are protected

17

by their inconspicuousness. As in those hen birds which, by similar means are rendered difficult to detect upon the nest, the different coloration of the two sexes indicates the greater importance of the female for the vital purpose of the perpetuation of the species.

Although having the same characteristics as horns, only a minority of the structures which, as I have shown, may be found upon any part of the body, can be described by that term. It is commonly applied to projections in the fore-part of the body either formed by the enlargement of the jaws, the mandible-horns, or by outgrowths from the upper surface of the head or thorax, the outgrowth-horns. These two categories include all the most fantastically developed forms of male accoutrement. The reason why it is in this situation only that such structures attain their greatest development is not difficult to appreciate. Projections from the lower surface of the body, from the legs or the wing-cases, so long as they remain of small size, do not seriously interfere with an insect's activity, but any considerable increase in size must result in hampering its movement and, if continued, at last impede it to such an extent as to threaten its extinction. The only exception is a backward extension of the wing-cases. These, as in *Diurus forcipatus,* Plate 4, figs. 5 & 6, may reach an extravagant length without causing more than slight inconvenience. Outgrowths of the upper surface of the head or thorax, however, may develop to an astonishing extent and still permit a considerable degree of agility and it is among beetles so equipped that we find the most extravagant forms. Such are the giants, the huge Hercules-beetles, (Plates 1a & 2) of South America, the Oriental Atlas-beetle (Plate 2), which aroused the admiration of DARWIN, and many more.

One other type of what we can only regard as extravagant development is found in the great elongation of some of the jointed appendages in certain male beetles. Being jointed, the insect is able to fold or move them in such a way that they do not impede it as a fixed outgrowth must do if greatly prolonged. I have already mentioned the enormous elongation of the antennae in the male weevil *Mecomastyx.* In the Timberman beetle, *Acanthocinus,* not uncommon in Britain, the delicate antennae may be between four and five times the length of the body and in some of the Anthribid beetles of tropical countries the antennae of the males reach a

18

still greater disproportionate length. Of a similar character is the grotesque elongation of the fore-legs found in a few giant beetles. The Harlequin-beetle, *Acrocinus,* so named from its curious decoration with black, white and red lines, is one of these. It is found in South America, in decaying Rubber-trees, upon the wood of which its grubs feed. The body is about three inches long, the legs slender, and the fore-legs of the female bear along the edge of the tibia a double row of very sharp spines, directed obliquely backward, which are no doubt of important assistance in tree-climbing. In the male these limbs are immensely elongated and may be more than six inches long. They are so jointed that they can be folded up and packed close to the sides but, even when so folded, project both in front and behind. It is strange to find that, accompanying this surprising lengthening, has gone a degeneration from the practical standpoint, for the spines have become reduced to mere vestiges.

Exactly similar is the case of a few quite unrelated beetles found in the East and also, like the Harlequin-beetle, giants. One of these, described and figured by A. R. WALLACE in his well-known book "The Malay Archipelago", is called *Euchirus longimanus.* It is a huge insect, more than three inches long, found in the Dutch East Indies, where it frequents the Sugar-palms, to enjoy the sweet exudation. The grub feeds, like that of the Harlequin-beetle, upon decaying wood, the female beetle laying her eggs in the moist substance, into which she forces her way by means of her strong, sharp-toothed fore-legs. The corresponding limbs of the male have lost all resemblance to those of the female and grown to a length about twice that of the insect itself. They are curiously toothed and twisted and, although capable of being packed together when the insect is at rest, are so unwieldy that it can only drag itself along slug-gishly by their means.

The analogy between these elongated legs and the greatly produced mandibles of many other male beetles is very apparent and in both cases we cannot fail to notice the apparent loss of efficiency in the change from the organs of the female to those of the male. The explanation is that the practical efficiency required from the female in order to ensure the perpetuation of the species is unnecessary to the male, which takes no part in the preparations for the future generation. Although, as in

19

Onthophagus sagittarius, lately mentioned, and other cases in which both sexes labour together in nest-building, we recognize special adaptations in the male for his particular share in the operations, in a greater number no such adaptation is apparent.

Of the two pairs of jaws possessed by beetles the maxillae are generally hidden within the mouth, while the mandibles are more or less exposed. In predacious forms they are used chiefly for seizing the prey and often project considerably, while in those that live upon vegetable food they are generally equally important as biting organs. In neither case are they liable to develop into horns. There are many beetles however which, during their brief adult life, take no solid food. The jaws of the females may be of importance to them for performing the operations necessary for the proper placing of their eggs but those of the males are without that employment. It is amongst such beetles that mandible-horns are often found and in some of them the jaws have developed to such an extent as to place them amongst the most fantastic of insects. It is in the Stag-beetles (Lucanidae) that the most remarkable development occurs and the largest beetle found in the British Islands, *Lucanus cervus,* Plate 8, figs. 1 & 2, is one, which sometimes causes alarm in London suburbs, where, although common enough, it is scarcely familiar, owing to the shortness of its period of activity.

There is one genus of Stag-beetles (*Nigidius*) of rather small size in which both sexes have processes upon the upper surface of the jaws, of rather antler-like aspect, and the unrelated tunnelling vine-cutter (*Lethrus*) has processes upon their lower surface in the male alone. In some species of the latter genus, very strangely, the processes on the two sides are not alike and in others they are absent altogether. (Certain little beetles belonging to the genus *Platydema* are known also in which the head bears two horns, one long and the other short). But it is by great elongation of the jaws that real mandible-horns are produced and this is a phenomenon peculiar to males. It is found in many of the Cerambycidae or Longicorn beetles, such as the fierce-looking but quite harmless South American *Psalidognathus* and the giant *Macrodontia,* Pl. 3, which may reach a length of seven inches from tip to tail. On a smaller scale many of the elongate Brenthidae (see Plate 4, figs 1 to 6), the females of which

20

have extremely minute jaws, exhibit a still more remarkable enlargement of the organs in the male.

Outgrowth-horns, whether arising from the head or thorax, are more varied in their origins than mandible-horns. Those upon the head may spring from its front margin, as in many of the Rose-beetles (Plate 12), from its sides, as in *Heliocopris gigas* (Plate 7, fig. 15) or from its hinder part (*Catharsius*, figs. 13, 14). They are extensions of pre-existing ridges, broadly based but tapering as they lengthen, sometimes pointed at the end, sometimes forked, knobbed or branched, sometimes finely toothed. The horns arising from the thorax also have various origins. Often a groove extends lengthwise across the middle thorax which may be replaced by a more or less extensive hollow and the fold of horny substance forming the boundary of this may be elevated in front, behind or on each side. The elevations may be short or form long slender horns. There may be a single horn, two lateral horns or all three together as in many species of *Strategus*. In the Indian *Peperonota* a single horn extends backward from the hind margin, in the Hercules-beetle, Plate 2, a massive dorsal horn projects straight forward, in *Golofa Porteri,* Plate 10, fig. 1, a remarkably long and slender one rises straight upward. In *Strategus Simson* (the trivial name is from the Swedish equivalent of Samson), shown on Plate 6, fig. 5, a Jamaican beetle, the two lateral horns are very slender and the median one forked at the end; in the equally curious Javan *Dipelicus Cantori* the long deep excavation is bordered by five horns, one upon the head, one on each side of the hollow and two behind it, producing the appearance of a capacious basket with its supporting framework projecting all round. The hollowing out of the thorax seems to reach its extreme limit in *Dipelicus Geryon,* Plate 13, fig. 4, in which the two posterior horns are replaced by a single sharp process.

The stages through which animals have arrived at the aspect they now present may sometimes be learnt by studying carefully the different species composing a single natural group. The great Elephant-beetles, forming the genus *Megasoma,* are instructive for the study of beetle-horns. The genus contains the largest of all known beetles, and indeed the most bulky of all insects. Not all of its eight different species are giants. Two of them are not much more than an inch long, while the largest, *M. elephas*

21

and *M. Actaeon,* are about 5 inches in length. The males of all have a head-horn which divides at the end into two prongs and may have a branch at its base. Some species have a horn rising from the middle of the back and some a pair of horns projecting from the front angles of the thorax. The size of all the horns is easily seen to vary in every species according to the size of the individual specimen and the average size of the head-horn in each species similarly increases according to the average size of the specimens composing the species; but this is not the case with the dorsal horn. This, which is present in the two smallest species, is absent in the two largest, in which it is represented by a very small vestige only. In two species of intermediate size a dorsal horn is present but the larger of the two species has a larger head-horn and a rather smaller dorsal horn. It is very interesting to note that the diminution of the dorsal horn is accompanied by a corresponding growth of lateral horns. As there is good reason to believe that the small forms are nearest to the ancestors of all and that the giants show the latest stage in the evolution of the genus, it seems that a gradual increase in size has taken place in past ages, accompanied by the enlargement of the horns upon both head and back, until a certain stage, although the head-horn continued to grow larger, that upon the back began to shrink, while, in compensation, lateral horns were produced. In the giant *M. elephas* these lateral horns are rather less developed than those of the equally gigantic *M. Actaeon* but the process at the base of the head-horn is larger. This balanced increase and decrease, revealing an automatic element in horn-development, is very significant. There is no reason to doubt that all the forms had a common origin and it seems possible that in the course of their evolution the dorsal horn reached a stage in which it became an impediment and began to diminish again, disappearing altogether in certain of the species. Obviously it was not necessary and it may well have become an encumbrance, while two or three small projections must be admitted to be less encumbrance than one large one. It has been shown that outgrowths of the same kind as horns may be found in almost any part of the horny exterior of a beetle and if the reduction or disappearance of an outgrowth entails an equivalent growth at another point this is to some extent explained. It becomes clear also that it is not

22

necessary to assume any definite use in order to account for the existence of any particular outgrowth. That some have acquired uses and become adapted for their better performance need not be doubted but such acquired uses do not account for their first appearance and still less for the extravagant degree of development shown by many of them. It will be found that it is only amongst forms of very moderate development that adaptations we can perceive to serve practical purposes occur. Of the rest, so immensely varied are they, so inconstant is their degree of development and so fantastic the extravagance exhibited by some of them that speculation as to how and why such forms have come into existence could not fail to be aroused. In the hope of gaining some light upon these problems I have attempted to bring together whatever reliable information can be obtained upon the subject of the habits of the insects concerned, as well as to ascertain by a closer study of the structures themselves what light is shed upon the problem of their significance.

CHAPTER 3.

THE HABITS OF HORNED BEETLES.

In order to understand the significance of horns, whether in beetles or other animals, it is essential to know as much as possible of the habits of their bearers. Those of the horned beetles, in their general outline, are like those of all other beetles, that is to say they begin life as fleshy grubs hatched from eggs formed in the ovaries of a female beetle and fertilized by a male. The eggs and young larvae being very delicate and liable to be destroyed by a great variety of adverse circumstances, their survival depends upon the precautions, differing to some extent in nearly every species, which the special instincts of the mother-beetle lead her to take and which may be of an astonishing complexity and require in some exceptional cases the collaboration of the male beetle. Marriage by capture is, according to all observation, the universal rule with beetles and, although DARWIN considered it possible that females could exercise choice amongst their suitors, no evidence has ever been found to controvert the statement that the male is the active and the female the passive sex. The sensory organs (eyes and antennae) of female beetles are very commonly poorly developed compared with those of the males and the antennae especially, the seat of the olfactory sense, are in a very large proportion of the males much better developed than those of the females, for the reason that the male invariably seeks out the female and not the contrary. Male beetles very frequently have more muscular legs than their females, sometimes provided with hooks and other contrivances for grasping and some of the outgrowths upon the legs which have been described here may perhaps serve as adjuncts for the same purpose.

But there seems no evidence of any adaptation of horns for the same purpose. The mandibles of many beetles are normally prehensile and perhaps sometimes employed in holding the females but their elongation reduces their gripping power in proportion to its degree and the great though strangely variable enlargement of these organs seen, for example, in the Stag-beetles, cannot increase their efficiency. Outgrowth-horns upon an insect's back are debarred by their situation from employment as grasping-organs but in some of their forms these can be seen to have become adapted for employment in the operations of nidification where, as mentioned above, the male insect bears his share in these activities. To what extent this occurs is not yet known but it is quite likely that collaboration between male and female beetles is more common than we have had reason to suspect. We can only explain a comparatively small number of the different horn-forms in this way and it is probable that in most cases there is no collaboration of male and female. The many extravagantly developed forms of horn remain unexplained and indeed appear to effectively debar their wearers from sharing the labours of the more workmanlike females.

It has already been indicated that the best examples of horned beetles and indeed the majority of all those known to us are to be found in the great group called, from the fan-like form of their antennae, the Lamellicorn beetles. These consist of three families, the Passalidae, one of the most peculiar of all groups of beetles, the Lucanidae or Stag-beetles, remarkable for the astonishing mandible-horns of their males, and the Scarabaeidae or Chafers, an immense family many times greater than the other two together, which contains all the most fantastic of those beetles with horns in the narrower sense of outgrowths from the fore-part of the body. In each of the three families individual cases have been found in which the two parents work together to provide for the necessities of their brood.

The Passalidae are found only in logs and tree-stumps in a decaying state, upon the substance of which they feed. The adult beetles and their larvae live together in galleries excavated in the rotting wood and observers in various parts of the world have recorded the discovery of parties consisting of the two parents and a variable number of young of different sizes. The evidence seems to show that the latter, at least in their early stage, are dependent upon food provided for them by their parents. Many of the Passalidae bear a short horn upon the head in both sexes but many others are without it.

The Lucanidae, the mandibles of whose males are so curiously trans-

25

formed into horns, have completely different habits, although their larvae also feed upon rotting wood. Arrived at maturity, the beetles leave the log or stump upon which they have lived hitherto, the male never to return and the female returning only to deposit her eggs. The larvae on emergence find abundant food awaiting them and may feed for several years before reaching their full growth but the two generations never meet. The only known exception is the peculiar genus *Sinodendron,* containing two species in the Old World, including Britain, and two in the New World. In these four, almost alone amongst the Lucanidae, the mandibles are not enlarged but there is a short horn upon the head and another, still shorter, upon the thorax; and, so far as we know, in *Sinodendron* alone in the Lucanidae, the two sexes in concert prepare a burrow and provision it for the young. The British species, *Sinodendron cylindricum,* is shown in Plate 7, figs 9-12.

The multitudinous Scarabaeidae show much greater variety, both in their form and manner of life, than the two other families. Many of them have a very simple life-history like that of the Stag-beetles, their larvae feeding upon rotting wood or the roots of plants; but others, including many of the horned forms, have habits of a much less simple kind. The mother-beetles in some cases, and both parents in collaboration in others, prepare a nest and store it with food for their progeny, often apportioning the right quantity for each and fabricating for them carefully shaped cells. These labours often occupy most of the summer and some of the beetles probably rear broods in successive years. For reasons which will be explained we must conclude that those male beetles with an exaggerated horn-development bear no part in these operations. As in the Stag-beetles with greatly enlarged mandibles in the male, it appears that among the Scarabaeidae also extreme dissimilarity between the two sexes is accompanied by a life-history of a very simple kind in which there is no collaboration between male and female.

When we attempt to discover the possible uses of horns by ascertaining in more precise detail the behaviour of particular species we meet with very serious obstacles. The study of insect bionomics is a very fascinating one but also one of great difficulty, needing for its success, not only great patience and perseverance, skill and resourcefulness, but also insight and

26

deductive capacity, together with the power of holding the imagination severely in check. It is not very surprising therefore that, compared with the vastness of the subject, the total amount of our information concerning it is small, for those competent to pursue it are few. For those blessed with the necessary qualities and opportunities for exercising them the possibilities are almost unlimited. The greatest work on the subject, the ten volumes of "Souvenirs Entomologiques", by the celebrated French writer, J. H. FABRE, dealing as they do almost exclusively with the insect inhabitants of a tiny corner of France, illustrate in a striking way the immensity of the territory still awaiting exploration in the rest of the world.

The published records dealing with the habits of horned beetles are not only few but of very unequal value. A single observation is an insufficient basis from which to deduce a habit but when the records of more than one observer confirm one another we are on safer ground. Conclusions, which may only be the first interpretations that come to mind, are sometimes drawn from single observations, perhaps made under conditions which did not permit absolute certainty as to what actually occurred, and these must be treated with every reserve. For instance, when we are told, on the authority of a distinguished German professor of a century ago, J. C. ILLIGER, that one of the ball-rolling beetles, having, in the course of its operations, lost its food-ball, which fell into a hollow from which, by its unaided effort, the insect was unable to retrieve it, the beetle went away and later returned with three companions by whose combined efforts the object was recovered, after which the three companions resumed their own occupations (MULSANT, Coléoptères de France, Lamellicornes, 1842, p. 42), we need not question the actual occurrence of the events reported; since, however, it is not stated that any distinguishing mark enabled the observer to make certain that the original owner was one of the four beetles which extricated the ball and the one which remained in possession of it at last, we are entitled to question whether the interpretation of the events is the true one. The facts as related seem rather to confirm other accounts that state that the food-ball is often the object of a contest in which either the real owner or another beetle may be the victor.

It is well also to remember that in describing the habits of insects, and

27

especially in estimating the objects and impulses behind their actions, the only terms which language provides are those applying primarily to animals of quite another, usually the human kind, and therefore, it may be, only applicable in a more or less inexact sense to creatures of a very different type. The mind of an insect cannot be adequately interpreted even by the imaginative genius of a MAETERLINCK. It is a great merit of FABRE'S work that he was accustomed to relate in detail what he actually saw, so that his interpretations, which other observers have not always confirmed, can be judged according to their merits.

The difficulties of investigation are especially great in the case of burrowing insects such as are most of those dealt with here. Their operations are performed in darkness and cease if a ray of light falls upon them. Much of our knowledge therefore must be a matter of inference from such evidence as can be found in the results of their operations. Such circumstantial evidence may be more important than reports of observed behaviour which may be quite differently interpreted by different observers. It is essential to realize that the imaginative and deductive faculties no less than the power of exact observation must be exercised in this enquiry. Comparatively few as are the insects of whose observed behaviour any exact record exists, the knowledge to be derived by deduction and generalization from the facts which have been securely established is considerable. Having ascertained the manner of life of a few insects having a particular type of structure in common we can safely infer a similar way of life for others with a similar structure. Thus if the habits of a beetle inhabiting Britain are known and another, different but of closely similar type, is found in Africa or America it is permissible to pronounce that its habits, however differing in detail, will be in the main the same. Although a knowledge of the living insect is indispensable as a basis, the comparative study of forms from all parts of the world and only to be brought together in the dead state is also important. It has been mentioned that most of the insects with which we are here concerned are burrowers. The implements with which their digging operations are performed are chiefly the fore-legs, which bear teeth at intervals along the outer edge. The teeth, at first sharp, become worn by the scraping process and may become entirely worn away. But in some cases few traces of wear can be

28

discovered and I have found that by careful attention to these teeth valuable evidence as to the habits can be obtained. Provided a sufficient number of specimens can be examined, it may become clear that both males and females in some cases and in others females alone bear upon them such evidence of industry. From this evidence the important conclusion has been drawn that those beetles, invariably males, which bear horns of fantastically exaggerated size and form usually show no bodily indications of wear, in marked contrast to the females.

The need for an attitude of caution in dealing with insect bionomics is illustrated by the case of the Passalidae. It has long been known that, unlike most insects, these beetles live together in all their stages. Having an opportunity, during a residence in South Brazil at the end of last century, DR OHAUS, an entomologist with experience previously gained in Germany, spent some time in trying to get a better knowledge of their habits. In order to study their transformations he brought home the larvae, with the rotting wood upon which they appeared to be feeding, but was surprised to find that they invariably died, although larvae of other kinds, with apparently similar requirements, prospered well. Noticing that he generally found them in small parties, each accompanied by two adult beetles, he then took home the whole assembly and found that they lived quite well in captivity. If the adults were removed however the larvae quickly died. The two adults proved to be always of different sexes and close examination satisfied him that the jaws of their young had no mechanism for masticating solid woody substance. He therefore concluded that the grubs were dependent upon food prepared for them by their parents. He also believed that parents and young were able to communicate by means of the vocal organs which both possess. Beetles of various kinds, but by no means all, have, like other insects, special organs for producing sound but in the Passalidae the structural changes which have given rise to these are so great that we may reasonably conclude the organs are of particular importance to them. The adult beetle has upon its back, near the end of the abdomen, a pair of rounded bosses which, under the microscope, are found to be covered with short erect spines, extremely hard and massed very closely together. The wings lying upon the back extend as far as these bosses, where they fold back. At the fold is a tough leathery patch which is also covered with spines of quite a different character, short, very hard and pointed. The wing lies in a hollow in the corresponding wing-cover, with the spiny area outside and by rapidly moving the abdomen the insect can cause the two bosses to rub the spiny patches, producing a fairly loud squeaking noise. In some of the species the wings serve no other purpose than this and have become reduced to short strips of hardened membrane, each with its scraper at the end. The grubs have still more peculiar vocal organs. They are strange looking creatures because, instead of six legs, they appear to have only four, upon which, however, they can walk much better than the ordinary Lamellicorn larva. Carefully examined through the microscope, each leg of the hinder pair is seen to have at its base a flat surface covered with exceedingly fine, sharp and hard parallel ridges. Lying upon these ridges can be detected a tiny object like the paw of a microscopic dog, with four or five sharp claws, which are able to pluck the fine ridges and make them vibrate. These little paws are the missing legs, transformed in this surprising way into instruments for sound-production. The third pair of legs in the Stag-beetles, as well as in the Dor-beetles (*Geotrupes*), also bear special spines and are used for the same purpose but have retained the same form as the other legs. The changes by which the vocal organs of the Passalidae in both stages have been perfected are so profound that it seems not altogether unreasonable to suppose that they may be used for communication between parents and young.

The incident which led OHAUS to that conclusion was the following. Breaking up a log in search of larvae of another kind, he came upon a family of Passalidae, consisting, as usual, of two adults and about half a dozen grubs. Not wishing to take these, he put them aside and continued his search, during which he heard continuous chirping from the Passalids. Before leaving the spot, he turned over a large piece of wood lying near and found beneath it the two parent beetles and four of their progeny, while two others were making for the same shelter over various obstacles, blind though all these grubs are, and guided, as he was convinced, by the cries of those already there. It is unfortunate that no resident in a warm region, in which alone Passalidae are to be found, has as yet repeated OHAUS's experiment in order either to confirm or refute the theory of

30

intercommunication. I have been told by a well known entomologist, Col. F. C. FRASER, that he has on various occasions found in India Passalid families consisting of two adults and a number of larvae and Prof. W. M. WHEELER, in his book, "Social life among the Insects" (1923, p. 27), states that his own observations, made in Central and South America, Trinidad and South Australia, confirm those of OHAUS. Unfortunately he records no details. In a somewhat lengthy review of the subject (Ueber die Biologie der Passalus-Käfer), R. HEYMONS, who found the insects in the same regions as OHAUS, has contended that there is no sufficient reason for the belief that parental care is really exercised. By an investigation of the contents of the alimentary canal he decided that the larvae were able to digest and assimilate woody tissue in a raw state, and not, as OHAUS believed, dependent on predigested food. Experiments with the North American *Popilius disjunctus (Passalus cornutus)* made by a group of students of Duke University, N. Carolina, and described by them in "The Ecology of *Passalus cornutus*" (A. S. PEARSE and others, Ecological Monographs, 1936, p. 455) led to the conclusion that, although well-grown larvae could be reared upon rotting wood, newly-hatched specimens needed food previously dealt with by the adults. When separated from their parents in an early stage they did not survive. We may conclude that the degree of dependence varies according to the species, of which about 500 are known in different parts of the world. OHAUS says that he found more than thirty different species, but he has not recorded the names of those to which his observations refer. The experiments of HEYMONS were made with *Passalus interstitialis*. This, as well as *Popilius disjunctus (Passalus cornutus)* and many others, bear upon the head a short stout horn directed forward, but none of those who have studied the insects have attempted to explain this. It is perhaps of use in clearing rubbish from the burrow. The horn of *P. disjunctus* is of very peculiar shape. It rises perpendicularly from the head just behind the middle and is abruptly bent at a right angle, extending forward nearly to the front margin of the head. Certain Passalidae, in most of the regions inhabited by them, are found between the loose bark and the wood of old trees or in cracks in the wood and, in order that they may be able to squeeze into such small spaces, are extremely flat. Horns, in such a case, would

31

obviously be in the way and they are always absent. In Australia and Malaya occur peculiarly convex forms belonging to the genus *Aulacocyclus* and it is interesting to find that, although these are less closely related to the horned *P. disjunctus* than are the flat forms, they bear a horn of the same peculiar hooked shape. Still more remarkable, another genus, *Phrenapates,* which belongs to a quite different family, the Tenebrionidae, and is structurally very different but has adopted a similar social mode of life, has a striking resemblance to the Passalidae and bears a similar horn. It is peculiar to Tropical America, and is narrow, black and shining, like the Passalids and of similar size, about an inch long. All the species have powerful jaws and a hooked horn upon the head precisely like that of *P. disjunctus.* About half a dozen species of *Phrenapates* are known and one of them was found by OHAUS in Ecuador. He found it living in smooth-walled galleries bored in the wood of the silk-cotton tree (*Bombax*). In these galleries were two parent beetles and larvae of various ages living together. The tunnel was about 18 inches long and had roomy niches on each side, at regular intervals, each occupied by eggs or larvae. The niches were provided with collections of loose woody fibres, upon which the larvae were feeding, and which he concluded were provided by the parents,

In all these cases the horn is possessed by both sexes alike and the occurence of an organ of this unusual form, together with habits of a similar kind seem to point to a similar employment.

It has been mentioned that the Stag-beetles (Lucanidae) are found, like the Passalidae, in rotting wood, upon which their larvae feed. The two groups differ in an almost startling manner in one respect. In the Passalidae, male and female are indistinguishable. Their horns, if present, and all other external features, are identical. In the Lucanidae, with a few exceptions, the two sexes are extremely dissimilar, so that it is often very difficult to associate male and female correctly. As a rule the males are recognizable by the want of compactness of shape, the elongation of their legs and still more of their mandibles, very evidently unfitting them for a narrowly confined life of the kind led by the Passalidae. In certain exceptional cases the two sexes are alike (e.g. the genera *Figulus* and *Nigidius*) and it would not be surprising if the habits of these, when

32

known, were found to be more like those of Passalids, but it is not possible to suppose that the males, at least of the majority of Lucanidae, can have similar habits. Although nothing appears to have been recorded about any except the very few European members of the group, such knowledge as we possess indicates very different conditions. The great Stag-beetle of Europe and the south of England (*Lucanus cervus,* Plate 8, figs. 1 & 2), may fairly be regarded as a representative species and its life-history presents a surprising contrast to those just recorded. Whereas the larval life of the Passalidae seems to be considerably shorter than the adult life, that of *Lucanus* lasts for three or four years and the active adult life for a few weeks only. The eggs are laid in the moist humus a little below the surface of the ground at the base of a decaying post or tree-stump and there left. The grubs feed upon the decaying woody substance, lying all the time upon the side of the body, until after many months they attain a length of about 3 inches. Each one then constructs with its jaws a cell of woody fragments, turns upon its back and casting its skin becomes an inert pupa, and a few weeks later a fully developed beetle. The final change takes place in the autumn but the insect remains imprisoned in its cell without nourishment of any kind until the early summer and, even after emerging into the outer air, does not appear to suffer from hunger. It merely sips a little liquid. What purpose then do the huge mandibles serve? Since, as in so many other insects, the transformations are undergone within a hard-walled cell constructed by the larva for the purpose and which must be broken before the adult beetle can reach the outer world, it might be supposed that its strong jaws would be used for this operation; but, although the cells of male and female are alike, their jaws are utterly different and the fact that, like *Copris* and related forms the jaws of which are quite soft, these beetles must wait for their release until the weather has so far relaxed the walls of their prison that they yield to the pressure of the unfolding body, indicates that this is not the use of the male horns. Are we to regard them as weapons? Contests between rival males of *Lucanus cervus* have often been seen. They push and struggle, open and close the mandibles and flourish them alarmingly but can cause no damage such as might result from the smaller but much stronger jaws of the female. There is a large Indian stag-beetle, *Hexarthrius*

33

Parryi, the males of which have mandibles much stouter than usual, with several sharp teeth upon the opposed edges which appear quite capable of inflicting injury. The examination of numerous specimens of this insect revealed various scratches and punctures upon the wing-cases, in all probability attributable to rival males, but none of a serious character. This species has an exceptionally formidable aspect however and I have examined large numbers of other Stag-beetles without finding a scratch upon their well-armoured bodies.

Not only are the elongated jaws of the male Stag-beetle less powerful than the short jaws of the female but the shape of his body and elongation of the legs show plainly that he has not her digging power. Although the legs may be toothed, the teeth are feeble and far apart, unlike those of the other sex. This applies to all the large species in all parts of the world and it may be concluded that the manner of life does not greatly differ from that of *Lucanus cervus.* The small forms often show no such dissimilarity between the sexes and these will no doubt be found to have more varied habits.

One of these smaller forms, *Sinodendron cylindricum* (Plate 7, figs. 9—12), has already been mentioned. It is not uncommon in Britain and, with three related species in other parts of the world, is remarkable for the fact that, unlike other Stag-beetles, the male, instead of elongated mandibles, has a short horn on the head and a shorter one on the thorax. All the front part of the thorax is hollowed out and the thoracic horn projects over the hollow from its hinder edge. The head-horn has a fringe of hairs on each side and the whole effect is to form a kind of scoop pointed in front. The female is without this but both sexes have the same cylindrical shape and rake-like fore-legs, with strong sharp teeth upon their tibiae, while the four hinder legs have close-set hairs and spines, forming stiff brushes directed backwards. The habits of these beetles were described by DR CHAPMAN in 1868 (*Ent. Month. Mag.,* 5, p. 139).

DR CHAPMAN found them in the process of nidification. A burrow about six inches long was driven into the dead and rotten wood of an old ash-tree by a pair working in collaboration. This was started sometimes by a male and sometimes by a female but soon afterwards a pair were always found at work together, the female extending the burrow, while

34

the male appeared to employ himself chiefly in removing the excavated material. Widenings of the burrow occurred at intervals, apparently to allow the insects to turn round. In branch tunnels eggs, 20 or more in number, were laid at regular intervals of about one-eighth of an inch in a spiral line round the wall, each in a slight depression, and the branch was afterwards packed with wood-dust. Each grub on hatching bored straight into the wood, the mother beetle remaining in the main burrow. According to DR CHAPMAN'S account "before any eggs were laid the female beetle was always at the side of the burrow, with her head at its extremity, as if continuing the excavation, and the male always had his head directed towards the opening, and often close to it, the remarkably flat front of his thorax nicely fitting the burrow and with sawdust sticking to it, as if it were used for pushing out frass." The presence of the male near the opening prevents the entrance of insects seeking to prey upon the eggs or young and it is probable that one or both parents usually die in the gallery, so that this protection is afforded until the brood are fully developed and able to gnaw their way out.

The whole structure of the insects can now be clearly understood as exactly adapted to their manner of life. The size and cylindrical shape of the body determines the form of the burrow, the beetle steadily gnawing at the wood, while slowly revolving all the time like a gimlet, the front legs raking away the loose material, which is passed back and swept to the rear by the four posterior legs, with their stiff brooms. When the female has taken on this part of the work and a quantity of wood-dust has accumulated behind her, the male removes it and ejects it from the mouth of the gallery, the distance it has to be transported continually increasing as the work proceeds. DR CHAPMAN found the length of the main gallery sometimes as much as 18 inches, but considered from 4 to 6 inches to be more usual. The distinctive shape of the fore part of the male is now explained. The rounded contour of the female, though best for the boring operation, is unsuitable for pushing a mass of loose débris through a tube. The hollowed front of the male is what is wanted for that purpose. This forms a scoop, of which the head is the lip, the short horn at the front margin, with its fringe of hairs, serving to collect stray particles, while the ragged ridge formed by the projections of the thorax at the hind margin of the scoop prevents the escape of any of its contents behind.

We can realize the convenience of this particular conformation still better by examining other wood-boring beetles, many of which show a more or less similar form but, as most of them have no means of turning round in their burrows, the shovelling must of necessity be performed by the hinder, instead of the anterior, end of the body. Another British insect of whose habits CHAPMAN has given a careful account is the Ambrosia-beetle *Platypus cylindrus*. Its body is cylindrical, like that of *Sinodendron*, but still narrower and much smaller, about a quarter of an inch long. Since its tunnels are of the same diameter in every part it cannot turn round but must move backwards in order to return to the entrance. Unlike *Sino-dendron* therefore, it cannot use the fore-part of its body in clearing the burrow but must do it in reverse. The wing-covers of the male, instead of being rounded as usual, are abruptly bent down towards the end, forming a flat surface, and at the fold each of the longitudinal ribs upon their back is slightly produced and carries a brush of stiff hairs, serving to prevent particles getting wedged between the insect's back and the wall of the tunnel. Each wing-case has also towards its end a little blunt process at the outer edge, which no doubt serves, like the frontal horn of the *Sinodendron,* to collect stray particles. The whole machinery is therefore the same, although formed of entirely different elements.

The two parents of *P. cylindrus,* like those of *Sinodendron,* are found with their brood and the borings are generally made in rather solid stumps of oak. When the main gallery has been driven about 6 or 7 inches, eggs are deposited at its end and a kind of seed-bed is apparently prepared upon which the tiny food-plant is already growing by the time the young larvae emerge. A branch gallery is then begun and the process repeated. The family may ultimately amount to from 60 to a hundred larvae of various ages. Parents and young live together all the winter. The workings grow more and more extensive but are always kept clear and smooth. According to CHAPMAN, "the principal function of the parent beetles, after oviposition, appears to be the ejection of frass from the open mouth of the gallery, which they alone appear to do. I have seen a small quantity brought every few minutes, at a season when the larvae were busily feeding. It seems to be done by the male or female beetle indifferently. I have strong reason to believe that either of these directs the movements of the larvae in the burrows, not only from the burrows containing eggs and young larvae being kept undisturbed, but also from larvae falling out of the burrows from which the parent beetles had been removed, a circumstance that does not otherwise occur." As it is in the male alone that the special flattening of the hinder end of the body and toothing around the flattened surface are found, it seems probable that DR CHAPMAN was mistaken in believing that the ejection of debris is performed indifferently by both sexes and this is confirmed by observations made upon a related species of the genus.

The manner of clearing out the galleries has been described by DR BEESON (*Indian Forest Records,* 6, 1917, p. 15) in the case of an Indian species, which, unlike *P. cylindrus,* adds to the entrance of its dwelling a short waxy tube serving as a bar to intruders. In this case "the whole of the excavation of the galleries is carried out by the female beetle and the male very rarely co-operates in this work, nor apparently does he assist in the care of the larvae and culture of the food fungus. He is invariably to be found within the first half-inch of the entrance tunnel, head directed inwards, occupied in protecting the entrance from the intrusion of enemies and in ejecting particles of frass and excrement. The method of ejection is somewhat striking and again illustrates that the peculiarities in the anatomical structure of this species are adaptations for a special function. The male beetle collects particles of wood-fibre, pellets of excrement and waste material and conveys them to the end of the entrance tunnel by pushing backwards with the hind end of the body. A small heap of the material is collected in the concavity at the end of the abdomen (formed by the dentate and setiferous edges of the elytra and the ridged terminal segment) and expelled from the wax tube with a sharp jerk. The muscular effort in ejection is very considerable and is often sufficient to throw small pellets several feet into the air. In consequence of this method of removing the frass, the waste material of the galleries is not accumulated directly at the base of the tree, but is scattered over the surrounding undergrowth in an area of several yards radius."

Species of the genus *Platypus* found in America have reached, according

37

to H. G. HUBBARD (Ambrosia-beetles. - *Yearbook of the Dept. of Agriculture,* Washington, 1896, p. 421) a more advanced social organization than their European representative. They form large communities and excavate a wide interior chamber to which rubbish is consigned, instead of expelling it altogether. They are consequently less restricted in their movements. According to HUBBARD "the older larvae assist in excavating the galleries, but they do not eat or swallow the wood. The larvae of all ages are surprisingly alert, active and intelligent. They exhibit curiosity, equally with the adults, and show evident regard for the eggs and very tender young, which are scattered at random through the passages and might easily be destroyed by them in their movements. If thrown into a panic the young larvae scurry away with an undulating movement of their bodies, but the older larvae will frequently stop at the nearest intersecting passageway to let the small fry pass and show fight to cover their retreat." HUBBARD also states that the males are savage fighters and the galleries are often strewn with the fragments of the vanquished. "The projecting spines at the end of the wing-cases are very effective weapons in these fights. With their aid a beetle attacked in the rear can make a good defence and frequently by a lucky stroke is able to dislocate the outstretched neck of his enemy." Having very carefully examined the ends of the wing-cases of *Platypus compositus,* the species especially studied by HUBBARD, as well as many other related kinds, I am unable to believe this possible. The processes are invariably blunt and clothed with bristly hairs. They are more or less divergent and often have a broad straight edge like a shovel. The broad, flat head can be instantaneously retracted and is then the least vulnerable part of the body. If disposed to fight, the very powerful fore-legs, with their sharp grappling hooks, used, as CHAPMAN has described, in tearing apart the wood-fibres, would be more efficient weapons. But it seems much more probable that the fragments of dead beetles found in the galleries were the evidences of an attack by predacious beetles of another kind, such as certain long-bodied Histeridae which are known to enter the runs of the Ambrosia beetles in order to prey upon them. HUBBARD'S account of the insects shows greater imagination, if less precision, than CHAPMAN'S a quarter of a century earlier. It is certain however that as yet we are only at the beginning of a knowledge of these insect communities. Whether future investigators will confirm in all respects the inferences of the early discoverers or not, the many different species, when their ways are brought to light, will undoubtedly reveal many fresh and unexpected varieties of social organization and behaviour. The much larger size of the family, as recorded by DR CHAPMAN for a pair of these beetles, than that found in the Lamellicorn beetles previously mentioned is a point of some interest. This contrast is equally conspicuous in the beetles which will be next described.

Although we have no evidence that the males of any of the typical Stag-beetles with exaggerated mandibles employ them in a useful manner or that any of them take a share in providing for their progeny, we should not be justified in concluding that mandibular horns are invariably useless. The remarkable mandibles of many male *Lethrus,* which have a horn on the lower surface of each, have been mentioned. One member of the genus (*Lethrus apterus*) is found in Eastern Europe, where it is known as the "Rebenschneider", or Vine-cutter, from its habit of cutting off the tender shoots of various plants, with a predilection for the vine. Many accounts of the "Rebenschneider's" habits have appeared in foreign journals. The best I have found is that of JACOB SCHREINER, in *Horae Soc. Ent. Rossicae,* 37, 1906, p. 197. The beetles live in pairs in burrows dug in the ground, the female spending most of her time below, while the male is seen guarding the entrance to the burrow or actively engaged in bringing provisions for the brood. When the first warm days cause the buds to swell, a spot is selected in firm sloping ground, where flooding need not be feared, and the pair set to work. The tunnel is driven obliquely into the ground for about six inches and then descends vertically for about a foot. At the sides of the vertical shaft six or eight oval chambers, each about the size of a pigeon's egg, are hollowed out, each destined to be the cradle of a single larva and packed with the juicy shoots. The male beetle climbs the vines, snips off the shoots with his strong jaws and returns with them one by one to his companion, who presses them firmly into the prepared cell. When this is quite full, it is closed with earth and a small egg-chamber, in which is placed a single egg, is formed just outside the partition. When this has been shut off, another food chamber is prepared and the male sets out again in search

39

of the necessary pròvision. After securing his booty, he descends the vine backwards with it in his jaws and draws it into the burrow in the same position. He defends his nest vigorously against intruders of his own species, who sometimes try to gain entrance. According to SCHREINER, who watched many of these encounters, the stranger is usually driven off without damage to either. The fore-legs seem to be chiefly used in these struggles. SCHREINER considers that the horny appendages beneath the mandibles are probably used in excavating the burrow or removing obstacles to its construction, but it seems likely that they may be of assistance in climbing, especially in the descent, when the jaws are employed in grasping the burden and therefore cannot otherwise assist. All the Geotrupinae, to which *Lethrus* belongs, are burrowers and their legs are formed for digging and running and not adapted in any way for climbing. Whether the other species of *Lethrus,* most of them inhabiting Western Asia, have the same climbing propensity as the "Rebenschneider" we have no information. Some of them are without the mandibular appendages and it would be interesting to know to what extent their manner of life differs from that of their European cousin. The grub of the "Rebenschneider", immediately it leaves the egg, finds the stored food and attacks it, in three weeks or a little more it is entirely eaten, he has attained his full size and changes to the pupal form. The final stage is reached about July but the beetle does not leave its underground cell until the following spring.

In the Geotrupinae, several of which are common in Britain, the males often bear horns upon the head and thorax. Like the Passalidae, both adults and young often possess an apparatus by which a high-pitched chirping note is produced, although none are known to live in communities like the former. Some of the Geotrupinae feed upon truffles and other subterranean fungi, but the Dor-beetles of our pastures are dung-feeders. Like the "Rebenschneider", they work together in pairs to provide for their offspring. One of them, *Geotrupes Typhaeus,* Plate 7, figs. 1 & 2, which can be found on sandy commons, has a very remarkable male and was the subject of one of the careful investigations of FABRE (*Souv. Ent.,* 10, p. 1) by whom it is called the "Minotaur". The name refers to the pair of horns, which are borne upon the thorax instead of the head.

40

There are actually three horns, the two outermost longer than the middle one. The thorax is very short and broad and on each side a long spike projects straight forward, a little flattened from side to side and bluntly pointed at the end. From the side these are seen to be directed slightly downward and between them, immediately above the head, a shorter but rather sharper horn points slightly upward. The insects are about ¾ inch long, short and broad, the female almost hemispherical, with only rudimentary processes on the thorax. They are able to fly, unlike *Lethrus*, which has no wings. Making their first appearance in the autumn, each one makes a vertical shaft, in ground frequented by sheep or rabbits, and in this it passes the winter. Almost before the winter is over, about February, they may be found associated in pairs and engaged in preparing a nest for their brood. The position of the shaft is always indicated by the mound of excavated earth ejected from its mouth. The amount of labour involved in the work is astonishing. The shaft may be dug to a depth of four or five feet or even more. As in the case of *Lethrus*, the female performs the digging, while the male carries away the excavated material. He seems first to compress it and then, placing himself beneath his load, hoists it slowly to the surface. This first stage of the work appears to occupy three weeks or more, the female remaining always at the bottom of the shaft, while the male continually repeats the toilsome ascent and descent. The necessary depth at last achieved, the work is changed. A smooth-walled chamber is prepared at the foot of the shaft, a single egg placed within it and the chamber shut off with a barrier of sand, upon the other side of which a plentiful supply of food is then stored for the future grub. The egg increases greatly in size before hatching, so that a clear space is necessary all round it, as well as a supply of air for the young insect. The transport of the food is the function of the male, who sets out in search of dry and hard rabbit- or sheep-droppings. His widely separated, downward-pointing horns may seem quite convenient for propelling a rounded object over the ground but, like *Lethrus*, his method appears to be to drag his burden backward. In this position he is able, by some mysterious faculty, to locate and identify the tiny aperture leading to his own dwelling, for the beetles live in fairly numerous colonies, and so to return continually with his spoils. A correspondent of "The Entomologist" who watched a specimen at work gave the following account of the procedure (G. B. CORBIN, *Entomologist,* 7, 1874, p. 132) - "In searching for the pellets the beetle invariably went in the same direction and on finding one it seemed to be recognized by an application of the palpi. The pellet was then seized by the two forelegs, the hooks and pointed projections — of which the use was very apparent — holding it firmly, whilst the pellet was further steadied by the head of the operator. In this position it was dragged, the beetle going backwards in exactly the same track as it had come in its search, and it was surprising to see how tenaciously it held on to its prize, for in returning it often fell backwards over some impediment or other, but the pellet in most cases was held firmly, although sometimes the beetle and pellet went tumbling over each other. If the beetle chanced to lose its burden it seldom searched for it again, but went off on the look out for another. Having accomplished the task of pulling some three or four pellets to the base of the mound around the tunnel, the next thing was to find the tunnel itself, which seemed to perplex the poor insect very much, for with the labour of collecting these pellets it apparently had lost the knowledge of the exact position of the tunnel, but after a longer or shorter search it was sure to find the desired aperture, when, thrusting its head and thorax therein, it would remain motionless for a few seconds, as if making sure that all things were right. It then proceeded to drag the pellets, one by one, up the side of the mound, and almost invariably the beetle fell backwards into the mouth of the tunnel with its load, which, being released, rolled instantly down the oblique gallery. This having been accomplished, the beetle would return to the mouth of its subterranean nursery and remain very quiet, as if resting from the fatigue of its previous exertions, before entering upon the labour of pulling up another pellet. I thus saw the creature dispose of eight or nine of these pellets and at last left it resting in the mouth of its retreat." It is of rather special interest to note from this careful narrative that the beetle always set out in the same direction and returned backwards by the same route by which he had gone. His horns were not used and his eyes, in his reversed position, could be of little or no service and so, inevitably, he often fell over obstacles but he apparently possessed the sense of direction, so mysterious to us, by which the wandering hive-bee returns to its own hive, the solitary burrowing bee to the tiny aperture of its particular subterranean nest and by which dogs and cats, transported sometimes hundreds of miles from their homes, have found their way back. By marking his specimens FABRE found that the couples, if separated, repeatedly came together again in their own home. How the male is able to recognize his own burrow and, dragging his burden, to direct his backward steps there and whether there is some subtle emanation from the female far below, in the complete darkness at the foot of the long shaft, by which she can be distinguished from all others, we can only speculate.

Another task awaits the pair of beetles as soon as a certain number of pellets have been conveyed into the nest. The material has now to be subdivided and packed closely into the terminal part of the shaft, where it is separated by a thin partition of earth from the chamber in which the egg is deposited. The shredding is performed by the male with his toothed fore-legs, the pellet, according to FABRE, being held by the points of the trident and the female receiving the fragments dropped from above and spreading and packing them into the place prepared. But, since the trident is fixed above the insect's head and the manner of articulation of the fore-legs only allows them to be used beneath the body, it is not easy to understand how the horns can be employed in this way. It is also rather doubtful whether the apparatus described by FABRE enabled him to see the insect's exact method of operation. I am afraid therefore that we must discard the explanation of the horn as for holding as well as for pushing the object. It is more likely that the jaws perform an important part in holding and dividing the pellets. Several other kinds of Dor-beetles, the males of which are entirely without horns, perform operations exactly similar, showing that horns are by no means necessary adjuncts to their outfit.

MR HUGH MAIN, who has devised a method of inducing the insects to make their nests between two sheets of glass, has given some pleasing details of their proceedings as he has observed them. "The male carries down the food and the female takes it from him at or near the bottom of the shaft, carries it in and rams it into the end of the chamber. She is very particular about the way the various layers of material are added and goes through a wonderful series of gymnastic feats in the process of packing them in... At times the male appears to get tired or perhaps lazy and does not return with a supply of material. The female then calls to him, using the stridulating apparatus situated on the base of the abdomen and the posterior coxae. It is quite audible to an observer. If he still lags, she comes up after him and gives him a good dressing down, clawing him vigorously, and he then once more resumes his task." (*Proc. S. Lond. Ent. Soc.,* 1917, pl. 18).

The preparation of a sufficient quantity of food for a single grub occupies several days. It is then enclosed and a lateral branch gallery is excavated above the completed section, another egg-chamber is formed, another egg deposited in it and a further supply of food brought and stored. The maximum number of the family thus provided for seems to be about eight. The earliest eggs are hatched before the last are laid, each grub devours the food provided for it and, when it is consumed, has reached its full growth. The young beetles emerge in the autumn and each appears to dig for itself a burrow in which to pass the winter.

FABRE has related an instance in which a female "Minotaur" refused to start domestic life with a particular male provided by him but at once accepted another. As a rule they appear to have been content with whatever matrimonial dispositions he made for them and this is the only recorded instance I have been able to discover in favour of DARWIN'S supposition that female insects exercise a choice of partner. No doubt the male in this case was incorrigibly lazy or otherwise deficient.

The stridulating or vocal apparatus mentioned by MR MAIN is found upon the hind legs. Near the base of each on the inner side, within the cavity in which it revolves, is a raised band of which the surface is crossed by microscopically fine but extremely hard ridges. By backward and forward movements of the leg these are drawn across a sharp edge within the cavity. This sharp edge plays the part of a violin-string while the moving ridges form the bow and the movements produce a similar result — a musical note, very soft and only audible to the human ear at a short distance. Although the vocal powers of the "Minotaur" are confined to the adults, it is a remarkable fact that other species of *Geotrupes* are vocal in the grub stage as well. The hind legs of the larvae, like those of the

44

adults, are used for this purpose, although the manner of using them is quite different and, unlike those of the adults, the hind-legs of the larvae are reduced in size and seem to serve no other purpose. A Spanish species, *Geotrupes momus,* shown in Plate 7, figs 3 & 4, very closely related to the "Minotaur" and with similar but still more curious horns, has also rather surprisingly lost the power of flight. There can be little doubt that the ancestors of these beetles were able both to fly in their adult stage and to squeak in the immature condition. Why some species have now lost these faculties we cannot say but it is evident that they have ceased to be of importance to them. When we survey the genus *Geotrupes* as a whole it is very apparent that the horned forms are much less abundant than those without horns, from which we may conclude that these accessories, if not actually a handicap to them, are at least not necessary for the proper performance of their functions. The horns are in every case peculiar to the male sex and, in some of the large exotic species, assume such grotesque forms that it seems almost certain any underground operation must be impossible for the males. For example, *Geotrupes Sharpi,* which inhabits Siam and South-West China, carries erect upon its back a long horn forked like the letter Y and on its head another, long, slender, curved and sharp-pointed, which passes between the two branches of the fork. It would be hardly possible to imagine an insect less adapted for burrowing than this. In the warm climate in which it lives it is of course not necessary, as it is for its European relatives, to burrow beneath the surface of the ground in order to escape the rigours of the winter but there can be no doubt that its helpless young, like those of all its tribe, live underground and are dependent upon a store of food provided for them there.

There is another genus related to *Geotrupes* and containing numerous species in many parts of the world, some of its finest representatives inhabiting Australia. This genus is especially remarkable for the variety of the horns borne by the males upon head and thorax, as well as for the smooth, globular shape of all its members. They are mostly yellow or rusty-red in colour and bear the name of *Bolboceras.* The burrows of one of them, *B. ferrugineum,* Plate 7, fig. 7, were found in North Carolina by MR A. M. MANEE. It is an insect about the same size as the "Minotaur" and it also works in couples. At certain seasons the beetles live singly in vertical shafts dug in sandy ground and the shafts are then always open at the top, but when a pair are working together the opening of the burrow is always closed by about an inch of earth, although its position is indicated by the mound of "sand-ropes" thrown out and lying on the top. Judging by the sketch made by MR MANEE (*Ent. News,* 19, 1908, p. 460), the quantity of sand removed indicates a burrow of considerable depth. The beetle seems able, by the exertion of enormous force, to eject a large quantity at a time, while still leaving enough to close the entrance. The work is in progress from early June to late August, but the discoverer never succeeded in finding the actual nest. He satisfied himself that nothing is carried into the shaft and we may conclude that the food of *B. ferrugineum* is similar to that of *B. gallicum,* which was found by FABRE to be certain fungi which exist beneath the soil, including the well-known truffle. The insects infallibly detect the presence of these by the most highly developed of insect senses, that of smell, and where they dig a fungus is always immediately below.

The smooth globular body which characterizes all the females of *Bolboceras* is very suitable for moving up and down a tubular vertical shaft but quite unsuited for removing from the shaft the earth or sand which must be brought up from below when it is necessary to penetrate to some depth. A glance at the male *B. ferrugineum* will at once reveal how this is accomplished. Across his thorax is a row of four short conical processes, joined by a transverse ridge, while his broad head is turned up at the front edge and the two front angles are drawn out to form two more upwardly directed points. As he ascends the shaft therefore his upper part forms a slightly hollowed surface enclosed between six little pointed elevations. As the material is scraped out by the female, the male places himself beneath it and, pressing against it with his head and shoulders, slowly climbs upward. When he has accumulated at the top as much as he is able to move at one time a final strong heave expels it, but he contrives always to leave at the pit mouth enough to bar the entrance to intruders.

The male of the French *B. gallicum,* a smaller species, has almost the same form as *B. ferrugineum,* but his head bears a single short stout horn

46

in front instead of two sharp angles. A common Indian species, *B. quadridens,* is still more like the American insect and many variations of the same form are found in different parts of the world, a feature common to them all being the short, but very sharp, points, usually four in number, projecting outward from the insect's back just behind the flattened area, of which two, either the inner or outer pair, are generally a little hooked. In *B. ferrugineum* the inner pair are widely separated and turned outward. In others all four points have a similar form. If, in imagination, we watch the beetle laboriously and persistently climbing his long steep shaft, carrying his burden upon his head, as he does week after week, we can realize that the possession of these grappling hooks must be of the greatest assistance to him and prevent many backward slips. His sharp claws and spiny legs are excellent for clinging, but the weight he carries and the force necessary for its expulsion need a hold upon more than one side of the upright tube. The grub of the Tiger-beetle lives in a vertical shaft, from which it suddenly attacks passing insects and drags them into the pit. It bears sharp hooks placed upon elevations on its back, which enable it to take a firm grip of the sides of the tube and there can be little doubt that the sharp projections, sometimes found only in the male *Bolboceras* and sometimes in both sexes, assist their operations in the same way. We can almost certainly explain as serving a similar purpose some strange forms in Tropical America, related to *Bolboceras* and with the same very short body. The male of one of them, *Athyreus Bilbergii,* has a sharp nasal horn rising from the front of his head and a stout two-pointed one projecting from the middle of his back. In another there are three points instead of two and in others the horn is shortened so that little more than these sharp points remain. All, no doubt, are aids in shaft-climbing and, since the females, in all cases, have only rudiments, we can conclude that the climbing is principally the business of the males, the females remaining below, occupied with more delicate operations.

But, if these strong sharp processes are obviously useful to their bearers, the very long and fragile ones occurring in other cases are as obviously of no such use. This applies to the only member of the group found in Great Britain, which may be considered the most notable British horned beetle, although only about the size of a cherry-stone. It is called *Bolboceras*

47

(or *Odontaeus*) *armiger* and has long been regarded as a rarity of mysterious habits, although is has been found in the neighbourhood of London. It is almost globular in shape and shining black. A well-developed male bears upon his head a very long and exceedingly slender horn and upon his thorax two sharp points, close together near the middle, and at the sides a pair of long lobes pointing backward, rather like a donkey's ears. These remarkable accessories render him as unfitted as could be imagined for such tasks as other males perform. Specimens captured have generally been seen on the wing and whence they came and whither they were bound has remained long an unsolved problem. Recently MR. C. J. SAUNDERS, writing in the *Entomologist's Monthly Magazine* (72, 1936, p. 178), has thrown light upon the mystery. A friend at Tilford, in Surrey, found that about ten minutes after the application of worm-killer solution to worm-casts on his lawn specimens of *Bolboceras armiger* in a comatose condition were often to be found beside the worm-holes, from which they had evidently been driven. Sometimes they were found in pairs but "the same burrow never produced both worm and beetle." It is clear that this insect escapes the necessity of excavating its own burrow, for which the male is so unsuitably equipped, by adopting a ready-made one. Can there be some connection between the extravagant development attained by the horns and the abandonment of the labours for which, in related species, they are found useful accessories?

A few other species closely resembling *Bolboceras armiger* are found in North America and no doubt have similar habits but what is their food, what provision they make for their progeny and what happens to the original owner of the burrow are still unsolved problems.

The vocal organs of some of the species of *Bolboceras* are exceptionally well developed. This is particularly the case in Australia, where some of the largest and most remarkable forms are to be found. It was first reported in 1865 that one of these insects was heard to produce a sound but the only subsequent account of the sound that I have discovered is in a note attached to a specimen in the British Museum captured by DR F. A. RODWAY in New South Wales, which states that this beetle "makes a noise like a puppy whining and can be heard across the room." The voice of the related European *Bolboceras gallicum* is also very audible and the singer has sometimes been discovered by its chirping, which FABRE describes as peculiarly sweet.

Another subfamily, the Scarabaeinae (Coprinae), related to the Geotrupinae, is even more remarkable for the horns displayed in immense variety; and the nesting habits of some of its members have attained a more advanced stage. To this group belong the Sacred Scarab and many other similar forms celebrated for thousands of years for their strange habit of making and rolling about a ball of food-substance, either for their own consumption or that of the young. Although these are not horn-bearers, very many members of the family are horned and some of them are amongst the strangest of all. All the Scarabaeinae seem to be scavengers, feeding only upon waste and decaying matter, and serving a useful purpose by burying it beneath the soil as provision for their young. Most of them are dung-feeders, some seek and bury carrion and a few have a preference for rotting fruit. They are very abundant in many parts of the world and, although mostly black, some display beautiful and brilliant colouring. They differ from the Geotrupinae chiefly in the structure of the head. In the latter this is comparatively small but very prominent and mobile, with the eyes, jaws and other organs projecting freely. In the Scarabaeinae the upper surface of the head forms a very broad flat plate, closely fitting into the thorax and almost concealing the eyes as well as the organs of the mouth.

The nesting instincts have attained an astonishing degree of elaboration in this group. They recall those of birds and indeed, in long continued devotion to the brood, few birds could compete for first place. The group is a very large one and very few of its members have yet been studied from the bionomic standpoint. Such knowledge as we have is chiefly due to FABRE. Happily a species found in Britain is one of those he has carefully investigated. This is *Copris lunaris,* Plate 5, figs 5 & 6, a shining black insect about three quarters of an inch long found in pasture land. The female is broadly oval and convex, with a transverse ridge upon the head where the male bears a strong erect horn. He has also a short conical horn on each side of his thorax and between these a broad double hump squared off perpendicularly in front and so, together with the head, forming a nearly round flattened plate upon which to lift the material dug out of

49

the burrow. The points of the short lateral horns form grappling-hooks and the hump between them is evidently derived from another pair of horns.

Each couple sets to work about April to establish the nest. Close to a patch of cow-dung a passage is dug obliquely to a depth of a few inches and at its end a large vaulted chamber, which may measure 6 inches in diameter and 2½ inches high is excavated and made firm and smooth. A quantity of dung is then brought into the chamber, the male collecting and transporting it little by little, while the female receives and compresses it into a uniform mass. A month or more seems to be occupied with this preliminary work. A sufficient quantity of material having been amassed, a portion of definite size is next separated from the mass and shaped into a ball by compression between the legs of the female. Whether the male takes any share in this part of the work is not known. The top of the ball is then pressed out so as to form a saucer-like hollow with a projecting rim. In this an egg is placed and the rim then drawn round it and the edges brought together above in such a way as to leave an air-space almost but not completely closed at the top. Room is thus provided for the swelling of the egg that takes place before hatching and a tiny passage is left for the admission of air to the larva. A pear-shaped object has now been produced and, when it has received the last careful shaping, another is formed in the same way. After three months' work about seven or eight such cells may have been constructed and each will now contain its young grub, feeding upon the interior substance. The work of the parents is not finished, however, for they continue to watch over their brood, repairing any breaches that may occur in the cells, suppressing the growth of moulds and providing against the various dangers to which the 'pears' are liable from the weather or other causes. FABRE found that if cells were removed and returned in bad condition to the nest they were speedily cleaned and restored by the insects.

The size of the *Copris* family is small. That of another species, *C. hispanus,* studied by FABRE is only half as many at that of *C. lunaris* or even less. These insects nevertheless are more numerous than many others that produce hundreds of young whose survival is left to chance.

Each larva consumes the food provided for it but leaves the outer crust of the cell intact. Within this it assumes first the pupal and then the adult form. Finally when the moist days of autumn soften the ground the insect bursts the crust of its cell and, after about five months' imprisonment, emerges into the open air. Winter is passed below ground again and in the spring the story is repeated. Whether parents and children ever actually meet seems to be uncertain, although from the small size of the brood it is likely that the mother often survives to produce a second or even a third. Is she conscious of the existence of the invisible young within the cell which she tends so carefully? Probably not. The bird brooding over its first batch of eggs can have no conception of the young ones which will in due time emerge from them, but the parental solicitude is not less on that account. The actions of beetle and bird alike are due to instinct; but is not the relation of the human mother to her newly-born infant also an instinctive one? Amongst our savage ancestors the infants of those whose maternal instinct was weak had small chance of survival and races deficient in that respect were inevitably eliminated. Whether in beetle, bird or man, continued existence is dependent upon a strong maternal instinct.

The length of life in the adult insect varies according to the magnitude of the tasks to be performed. With most insects nidification is a fairly simple matter and the adult life is correspondingly short; in some the operations of the female are lengthy and arduous but the male takes no part in them and lives a much shorter time than the female ; and in others again both share in the labour and enjoy long lives. Feeding habits have little to do with the length of life, for the parent *Copris* beetles, according to FABRE, whilst engaged in their long-continued operations underground take little or no food. It seems that abundant occupation results in longevity and death is due to the want of it.

Some Scarabaeinae, like the large species of the genus *Heliocopris,* Plate 7, figs. 15 & 16, instead of pear-shaped cells, make them perfectly spherical. These are enclosed in an outer crust of hard clay, half an inch or more thick and the egg-chamber is contained in this crust, always with an air passage leading to the exterior. Some of these balls have been found at a depth of six or eight feet below the surface of the ground. They may be four inches in diameter and are so hard that they have been taken for ancient stone cannon-balls. The South American *Phanaeus Milon* makes a similar cell with a thick crust of clay, which FABRE found to be provisioned with food derived from the carcase of a bird. The exact symmetry of the cells is always attended to with care and the marks left by the fore-feet of the parents in the process of moulding can generally be seen on the outside. How many are made, whether by one or both parents and by what processes, is known only in the European species.

It is certain that both parents do not in all cases share the work of nidification. FABRE has ascertained this in the case of *Onthophagus taurus,* a British species, so called from the pair of long diverging horns on the head of the male, which however are not like those of a bull in strength, but are very slender and fragile. They spring from the back of the head and curve outward and backward, embracing the thorax. It might be supposed that they would often be broken, but large series of collected specimens show that such accidents are very rare, and the reason seems to be that they are not employed for any purpose at all. The female alone excavates the nest and provisions it, after placing an egg in a roomy air-chamber.

The genus *Onthophagus* is an enormous one, containing more than 1500 species, most of them horned, and amongst them are some of the most extraordinary of all beetles in the extravagance of their armament. There is no doubt great variety in their life histories, but we know practically nothing on the subject except from FABRE, who, from his experience of half a dozen different species, pronounces that "the *Onthophagus* tribe know nothing of domestic ties." This generalisation is a rather rash one and probably untrue.

There is a kind of evidence which seems to afford some indication whether the work is shared by both sexes or not. The strenuous exertions made by many of the insects often leave their mark upon the implements employed, the most important of which are the toothed fore-legs. In fresh young specimens the teeth are quite sharp and, if the substance into which they burrow is quite soft, may remain so; but often the teeth are found to be blunted and they may even be entirely worn away. *The presence or absence of wear in these teeth therefore affords some indication of the insect's activities.* The examination of a few specimens only is of

52

little use, for they may show no wear because they have not lived long enough to acquire its traces; they may show it because of some exceptional difficulty met with on their first emergence from the larval cell. But if a considerable number of each sex can be examined and a distinct difference is found between the two sexes in the total amount of wear, this has obviously some significance. The great collections of the British Museum, which have been accumulating since SIR HANS SLOANE, in the days of Queen Anne, began to amass his "curios", provide long series of many of the species. As test cases we can take species of which the facts are known. I have been able to examine 49 males and 37 females of *Onthophagus taurus,* of which we learn from FABRE that the males do not co-operate in the labour of nidification. Of the 49 males, 16, i.e., less than one-third, show any wear, and of the 37 females 22 are worn, or nearly two-thirds. Of *Copris lunaris,* of which we know the two sexes to work together so long and arduously, I have examined 31 males and 24 females. Of these only 5 males and 4 females showed little or no wear, i.e., exactly the same proportion of each.

An abundant African beetle of the same group, *Catharsius tricornutus,* with three short horns in the male alone, one on the head and two on the thorax, may with almost equal certainly be pronounced a worker in double harness. Twenty males have the tibiae worn and 3 only unworn, 24 females have them worn and 5 unworn.

In a large African *Onthophagus, O. gibbiramus,* Plate 5, figs. 7 & 8, the male of which has a pair of very slender horns, still longer than those of *O. taurus* while the female has none, 10 out of 12 males are unworn and only 3 out of 7 females, evidently showing that, like *O. taurus,* the male is not a worker. In an Indian species, *O bonasus,* of which the two sexes are alike, both having a pair of fairly long but quite stout horns, 21 males out of 30, and 10 females out of 13, have worn front tibiae, i.e., practically the same proportion. It seems that here FABRE'S generalization does not apply and that the two sexes work together. Another remarkable African species, *O. Blanchardi,* has, in the male, a single slender thoracic horn, extending straight forward far beyond the head, which must surely be an obstacle to any serious work. 31 males of this species have all perfectly unworn tibiae. Of 38 specimens of the hornless female 23 show wear.

53

Onthophagus catta is another abundant species, found in Africa and Asia. The male has a pair of short horns, while the female is unarmed. Of 49 males and 52 females, 30 males and 33 females show evident effects of wear. This is exactly the same proportion of each and here again we seem to be entitled to suppose that the two sexes work together. A careful examination of many hundreds of examples, belonging to about thirty different kinds of horned species, led me to the conclusion that, *when the two sexes do not greatly differ or the horns of the male are of moderate size only, the fore-legs of both generally show the effects of use, in a similar degree; but where the male has an exaggerated horn-development, the evidence points to the labour being performed only by the female.* Either the existence of appendages of an embarrassing kind has served to prevent the male acquiring domestic accomplishments or the non-acquisition of the accomplishments has resulted in exaggerated development of the appendages.

The two greatest horn-bearing groups of beetles are the Scarabaeinae and the Dynastinae. The later sub-family includes the most gigantic forms of all. It is unfortunate that these are mainly inhabitants of hot climates and, since very few are found in Europe, very little is known about their habits except that, like the Scarabaeinae, they generally feed upon decaying matter. Many are found in rotting wood and some are found at work in couples. DR OHAUS has recorded (*Stett. Ent. Z.,* 1900, p. 215) that he found three different species, *Trioplus cylindricus, Phileurus latipennis* and *P. foveicollis,* living in couples, together with their eggs and larvae, in the interior of tree-trunks in South America. Each of these three species has a pair of horns upon the head, although in the last they are rudimentary. In the other two they are present in both sexes, but larger in the male. They stand up on each side of the head, like a pair of ears, blunt and rather flat, the head, in profile, looking rather like that of a little mouse. In *Trioplus cylindricus* we meet again a structure almost the same as that of the Stag-beetle, *Sinodendron,* the same narrow cylindrical shape and the same flattening of the fore part of the thorax, with a row of little elevations at the upper edge. Here again then, in another group of beetles, is a male whose business it is to keep clear of rubbish the galleries inhabited by his family. Do the short blunt horns serve to scrape together the particles of dust? It seems quite possible.

A related beetle found in Borneo, *Clyster Itys,* Plate 7, figs. 5 & 6, shows the same structure still better developed. The male is almost exactly cylindrical and very smooth and the front part has the appearance of having been cut transversely so as to leave four little elevations, which are evidently the remains of the four ancestral horns, in addition to which the head bears a single short horn. The female is quite differently shaped and we may infer that the division of labour is complete. Although nothing has been recorded of the habits of the species, we can tell that it is accustomed to bore tunnels in hard wood or soil, for the teeth upon the fore-legs are often distinctly worn down in males and females alike. This is not usually the case with members of its group, the Dynastinae, which seem rarely to perform tasks requiring such severe and long-continued labours as those of many Scarabaeinae. The jaws of the Dynastinae are much harder and stronger than those of the latter and are perhaps used to some extent instead of the fore-legs.

Like many Geotrupinae and Scarabaeinae, the Dynastinae generally have an apparatus which produces sounds, compared to the cheeping of nestling birds. It consists of a series of microscopically fine ridges near the end of the abdomen, which are made to vibrate by rubbing them against the posterior edges of the wing-covers. The vocal organs of the Scarabaeinae are of different kinds.

A well known Dynastine found in Argentina is *Diloboderus Abderus,* which is said to be so abundant as to cause damage to the grazing land in which it burrows. Its proceedings have been described by JUDULIEN (*Rev. Mus. La Plata,* 9, 1899, p. 371) and by DAGUERRE (*Rev. Soc. Ent. Agent.,* 3, 1931, p. 253). The beetles are about an inch long and live in pairs. The female is a dark chocolate colour, rather shining, oval and flattish, the male ashy-grey and very dull. His head bears a very long horn, curving backward and pointed at the end, and his thorax carries another, very massive, directed forward and forked at the end. This dorsal horn is hollowed beneath and the hollow is filled by a thick pad of velvety yellow hair. When the head is drawn back, the end of its horn rests between the forks of the dorsal one. Male and female are so entirely dissimilar that, but for the fact that they are always found together, they would be taken for insects of entirely different kinds.

The nest is an oval chamber about 2½ inches long and 1¼ wide, at the end of an oblique tunnel about a foot long. It is packed with fragments of leaves, grass, etc., amongst which about half a dozen eggs are distributed. Probably, after consuming the stored food, the larvae proceed to attack the living roots they find in the surrounding soil. OHAUS found that the larvae of other horned Dynastinae living in forest country, *Enema, Heterogomphus, Coelosis*, etc., live in smooth-walled burrows, from which they protrude their heads to feed upon rotting wood lying upon the ground above them (*Stett. Ent. Z., 1900*, p. 213). At the foot of the burrow was a chamber into which they quickly withdrew when alarmed. No doubt the burrow is the work of one or both parents. Whether the male *Diloboderus* shares in the work of nidification is not recorded. They have evidently not been observed to do so and the great development of the horns seems to make it improbable. They are apparently active in the daytime and energetically defend the nest against other males. After driving off his rival, the victor, according to DAGUERRE, expresses his satisfaction by a vigorous chirping. In fighting, the beetles are said to try to grip each other between the horns of head and thorax and two males have sometimes been found interlocked in this way, seemingly having been unable to disentangle themselves.

The big Brazilian *Enema Pan*, two inches or more long, has still longer but rather similarly formed horns on head and thorax in the male and Ohaus considered these to serve as defensive weapons against his foes, "even those of large size". He kept a male under observation for some time and says "I observed that the armature on the head and thorax serves not only to hold the female, but is also a good weapon with which it (the male) can protect itself against big enemies. I let the animal crawl round the table and, when he was about to fall over the edge, I seized the thoracic process with thumb and forefinger in order to bring him back, but I had hardly grasped him when he snapped the head-horn back and brought it with such force against the fork of the thoracic horn that the skin of my forefinger, near the nail, was broken and bled freely. He gripped a pencil in this way so firmly that I could lift him by it. If I approached him cautiously from in front with the pencil he lowered his head, with its long horn, like a bull, to draw it back when I brought the pencil near his back. If I approached him (? touched him) from behind, he raised himself on his forelegs and bent his head and thorax far back. All these movements are made with a liveliness I did not expect from so bulky an animal."

It may be supposed that the enemies of a beetle of this large size in Brazil would be birds of prey, monkeys and other insectivorous mammals and large reptiles. It is of course possible that it may sometimes defend itself as OHAUS describes, but there are reasons for doubting whether the horns are of any real importance for this purpose. OHAUS' description evidently refers to a specimen of large size, but, not only are the horns, as usual, liable to great reduction, so that the thoracic horn may present nothing to grasp, but *Enema Pan* is remarkable as an insect of which there are two very distinct male phases. The specimen referred to by OHAUS had a sharp-pointed horn upon the head and a strong forked horn on the thorax. In the second phase the thoracic horn is long and sharp-pointed and the one upon the head is still longer, with two cusps at the end. The female again has a sharp-pointed horn upon her head, but the thoracic horn is absent and represented only by a slight vestige. One only of these three phases, and that only in its full development, could be really efficient in the way suggested, whereas, if this were the real purpose of the horns, we should expect them to have been standardized at the stage of greatest efficiency. In insects, as in birds, we commonly find protective devices in the female which are absent in the male, the former being the more important for the perpetuation of the species, and it would be strange for a valuable means of defence possessed by the male to be absent in the more important sex. The extreme inconstancy of the male horns makes it equally improbable that their purpose is to hold the female.

There is one other of these beetles which has been seen at work. MANEE has described and sketched the nest of *Strategus Antaeus* (*Ent. News,* 19, 1908, p. 286). This is a shiny black insect an inch or more long, the male of which has three long horns on the thorax, one above the head and two behind, all the tips curving towards the middle. All three horns being immovable, there can be no gripping movements. The

57

5

account of its proceedings is as follows — "On the night of July 11th, 1906, I took my first females by electric light. That same month we investigated an inch hole by a cart-path and dug out a working male. On July 6th I took my first pair from between two exposed roots of a large oak. They were pulverizing the soil preparatory of shaft-digging. After several such takings of pairs and singles, I came to know the peculiar mound of earth, always pulverized to a depth of 1-3 inches.... Beneath the mound of loosened soil an inch shaft extends vertically for six or eight inches. At bottom of shaft a 1¼ inch chamber reaches horizontally for 1 to 5 inches and in this chamber, packed with finely broken bits of decayed oak leaves, a solitary egg is deposited. Sometimes two or rarely three such chambers diverge from the same shaft but, I believe, with never more than one egg in each. A favourite haunt for nesting is by a pile of dead oak-leaves, wind-blown in some hollow, from which I conclude that the young larva feeds on leaf-debris and later on decadent oak-roots." The digging fore-legs of this beetle are provided with very sharp teeth and males show the wearing down due to use to an even greater extent than females, from which we may suppose that they perform a large part of the work of excavation. This is certainly surprising when we consider the rather long horns some of them bear upon the thorax. It is true that these are fixed, with their points all turned inwards, while the head is unencumbered. There is a cavity between the three horns but no broad flat surface upon which to raise the dug-out soil, as in so many other excavators. It would be of the greatest interest to know how this is managed. There are many species of *Strategus,* most of the males having rather stout horns, with a deep cavity between them, but some have no horns at all. It is probable that all have similar habits.

The males and females of the nearly related Old World genus *Oryctes* seem also to work together, judging by their legs. A very well known species is the Black Coconut beetle or Rhinoceros beetle (*Oryctes rhino-ceros*), of which both sexes carry a single horn upon the head. This is said to serve a useful purpose by PROF. R. W. DOANE of California. I have not seen his original statement but reference to it appears in a German journal of which the following is a translation. "DR DOANE has made observations upon a horned beetle in Samoa which he has communicated to the weekly paper "Science". This beetle has a horn in both sexes but it is generally longer in the males. As there are males with a quite short and females with a long horn the determination of the sex by this means is not possible. The length of the horn varies between 1½ and 10 millimeters. Amongst mankind the beetles have a very bad reputation for they are destroyers of the Coconut-palm, searching out and biting the young leaves at its crown. Is has been found that in this deleterious operation the beetle employs its horn with great dexterity. It forces the whole horn into the substance of the plant and employs it as a kind of anchor whilst working with its legs or tearing the young leaf-tissue with its strong jaws." (*Entom. Zschr.*, Frankfurt, 27, 1914, p. 306). Whether this account of the insects' procedure can be confirmed remains to be seen. Their precise method of burrowing into the crown of a palm-tree must be very difficult to ascertain but the fact that both sexes perform the operation and both are horned adds probability to the explanation. In the *Philippine Journal of Science*, 1906, p. 143, Mr C. S. Baker states "Observation has shown that the males make burrows as well as the females and it is probable that they always accompany the latter at the time of egg-laying, retreating from the burrow they have made to allow the female access." The grubs emerging from the eggs so laid continue the work of destruction. But this habit of attacking palm-trees seems to be an acquired one, not common to the whole race, for the grubs are often found in quite different situations. They will flourish in rotting wood, decaying leaves and vegetable refuse and the original habit was no doubt to deposit the eggs beneath the ground.

The giant members of this same group of which the males exhibit the most extravagantly developed horns, the Hercules- and Elephant-beetles, etc. (*Chalcosoma, Megasoma, Dynastes, Golofa*), although the females are provided with digging fore-legs, are without these in the male. No observation of the living insects is necessary therefore to inform us that the habits are different. Mr A. H. Manee has stated that *Dynastes Tityus* occurs in large numbers in North Carolina, where it is attracted by fallen and rotting fruit of various kinds. He records that a boy took from a single ash-tree 31 specimens one day, 155 the next day and 189 on a third visit, while he himself found 387 specimens upon a single

59

tree. This insect has strange irregular brown blotches upon a dull green surface which give a curious resemblance to a decaying fruit and probably form a useful disguise, but, in many of the related forms, while the females have a rough, dark-coloured exterior, which makes them quite inconspicuous, the males are glossy and very conspicuous. The males of *Golofa* (Plate 10), for example, are shining red or orange and the females dull and black, the glossy green *Chalcosoma* has a dull inconspicuous female, as have the Hercules-beetle and the common *Dynastes Gideon* (Plate 8, figs. 3 & 4) of India and the East. Similar differences are not found in those that work in couples, where conspicuousness in the male might draw undesirable attention to both partners.

In the Oriental *Dynastes Gideon* the custom of marriage by capture has been observed in Java. The males of this have a strong horn upon the head another upon the thorax and both are forked at the end. By movements of the head the tips are separated or brought together and the insects seem to use their horns like a pair of tongs. The following account has been given by BATESON & BRINDLEY (*Proc. Zool. Soc.,* 1892 p. 590). "In view of the circumstance that there are scarcely any observations as to the functions of the horns of beetles, the following statements of BARON VON HUGEL are especially noteworthy. He says that the animals (*D. Gideon*) were caught by himself and by natives and were tied up with pieces of bast. When they were brought home and untied the males immediately sought out the females and, seizing them transversely, carried them about, held between the two horns, with evident satisfaction. He tells us that this was observed again and again and was clearly a definite habit. The males with small horns, though unable to lift the females, nevertheless made ludicrous efforts to do so. The habit is not confined to *D. Gideon* for BARON VON HUGEL observed it also on one occasion in the case of *Chalcosoma Atlas,* the well known Atlas-beetle." It seems that the acquisition of this habit has resulted from the form of the horns and not *vice versa,* for to many male specimens of *D. Gideon* and to all those of other species very closely related to it such a proceeding is quite impossible. In the same island of Java is found another *Dynastes* (*D. inarmatus*) scarcely differing from *D. Gideon* except that the male has a horn upon the head but none upon the thorax, while in the Punjab

60

another has only a vestige of a thoracic horn. In neither is gripping possible. Although well developed males of the Atlas-beetle have two long horns upon the thorax and one upon the head, gripping with them must be much more difficult than it is for *D. Gideon* on account of the way in which the head-horn curves over the back.

Another use of the horns has been described in one of the Elephant-beetles and in a species of the genus *Strategus*. MR WILLIAM BEEBE, of the New York Zoological Society, has watched males of the two species fighting for a female and has published a detailed account of their method. This is his account of the fighting of *Megasoma Actaeon* (the photographs show this species and not *M. elephas*, as it was supposed to be). "The opponents meet head on and either warily wait for the other to attack or one may rush headlong and begin the encounter. Usually both wait and spar at a little distance. The object first noticeable is an attempt with one or both fore tarsi and claws to trip and unbalance the opponent. This is evident in a long series of single photographs and in several complete kodachrome motion picture sequences. There are quick forward lunges and reachings out with one or both legs, sometimes at the same moment by both insects. This may or may not succeed but one will force the fighting and the result may be straight pushing and butting for a considerable period, exactly like two antlered deer. Now and then an effort will be noticed to lower the head and get the cephalic horn beneath the other insect. Again and again this is tried and both may attempt it at the same moment. Then recur the rearing and tripping attempts. Periods of rest or waiting may intersperse the encounter and twice I have seen one beetle turn and rush after the female. In both cases the other was after him full speed and the battle began again. The female never remained but went off as far as the confines of the cage or, in the case of the fighting taking place in the open, as far as we would allow her to go, when we would recapture her for fear of losing her in the under-brush. The only certainty was that she showed not the remotest interest in the encounter or in either of her suitors. . . . . The insects are generally horizontal when they begin pushing against each other but attempts at tripping will cause both to rear up high on the second and third pairs of legs. Then, if at all, comes the final phase, the all-out attempt to get

61

the tip of the curved bifurcated horn caught in the soft skin of the ventral joint between the thorax and abdomen. Once secured, we realize this is evidently the chief object of the encounter. The successful one puts forth all the strength of which he is capable and lifts again and again with all his might. The higher the other is lifted the more helpless he becomes as his feet, one after the other, leave the ground and with several super-beetle flings the victim of this grip is thrown over on to his back. Not once, but again and again this was the end result. Often the beetle simply rolled over and came back on his feet again and the whole engagement recommenced but sometimes he landed on his back and if the surface was at all level and smooth he spun helplessly, waving all six feet in mid-air. The winner began searching in all directions, evidently for the female." (*Zoologica*, 29, 1944, p. 55).

The word "evidently" was necessary here because the female had disappeared, so that the struggle led to no result.

MR BEEBE'S description of the fighting of *Strategus Aloeus* is as follows - "*Strategus Aloeus* fought as readily as their larger relations and in almost exactly the same manner. An important difference between the two species is that *Strategus* has all three horns on the thorax, while *Megasoma* has the central curved horn on the head itself. Although thus denied the inter-mobility of the horns, the smaller beetles fought with equal fury and quite as satisfactory results. The general plan of battle was identical, to get the anterior horn beneath their opponent and lever him up and over. In one of the first fights watched there were several momentary lockings and once the attacker was himself pried into the air and almost on his back. When upside down this species seemed almost more helpless than the larger and I believe would die of starvation on a smooth surface if left to themselves. The general movement and activity was less evident, owing to the lack of separate play of horn number one, but there was no lack of fierce effort."

It is to be noted that the weapons used in the two cases are quite different. In both there are three horns but one alone seems to be employed, borne in *Megasoma* upon the head and in *Strategus* upon the thorax. In very large specimens of *Strategus Aloeus* this horn extends a short distance beyond the head but in most not past its base. Indeed, in half the males I have examined, it is only a very small pointed projection behind the head. I have already mentioned that several species of *Strategus* have no horns at all and, as the males have probably all the same fighting propensity, it is likely that they manage equally well whether horned or not. If, in some of them, a horn is so situated as to come into play — well and good, but if not it is no matter. MR BEEBE remarks, of a contest between a large and a small *Megasoma,* that the latter's "very smallness of size was a help in some ways". This was no doubt because the head-horn, being much shorter in a small male, was more manageable.

Other supposed uses of beetle-horns have often been put forward but nearly always without evidence of sufficiently close observation or the necessary verification by repetition; some are mere folklore passed on locally from generation to generation without any attempt at verification.

In *Bull. Brooklyn Ent. Soc.,* 7, 1885, p. 221, it is said of *Dynastes Granti* in Colorado "They are always found near the tips of branches (of the mountain-ash) where, by means of their projecting thoracic horn, they scrape through the soft bark to cause a flow of sap, which is very sweet, and of this consists their food." This species, like others of its genus, has a horn upon the head directed upwards and one upon the thorax pointing forwards but a very slight examination is sufficient to show that it is almost impossible that either could serve the purpose stated.

For such an operation the strong mandibles are the obvious instruments but a casual observer might easily suppose that the horn was being used. A more circumstantial account by DR A. L. BENNETT, in *Proc. Ent. Soc. Lond.,* 1899, p. 11, concerns the West African Goliath-beetle *Goliathus Druryi.* "In November 1897, being then in the Bulu country, I started up the side of a mountain with a native guide to collect monkeys.... While taking a short rest during the climb, my attention was arrested by a sound overhead not unlike that of steam escaping from a small safety-valve. The noise was made by a large beetle which was soaring around a large vine hanging from an immense tree.... Seated on a fallen tree, I watched it soaring round and round, giving forth the strange sound that first arrested my attention. It was an interesting sight. Rays of light penetrating the forest foliage caused the large wings to glitter and scintillate with a most beautiful greenish lustre. The insect finally paused in its circling and settled upon the vine high up and out of reach.... The beetle seemed to be digging away a portion of the bark from the vine and to be feeding. I cut away some of the bark and a white milky juice not unlike that obtained from the rubber-vine trickled forth. The beetle worked hard, small portions of bark fell steadily from above and soon the breach it had made was clearly visible. Shortly after a smaller beetle appeared, evidently a female, and the male who had been at work on the vine gave place to the new arrival. The female was soon busy at work and seemed to be engaged in abstracting the juice from the vine. My Bulu guide contrived to capture in a very clever manner the largest of the two insects which to my surprise and delight proved to be a perfect specimen of *Goliathus Druryi* (male).... During my residence in the Congo français I had further opportunity on three different occasions of observing these insects and watching them feed. At one time I was able to stand within a few feet of a male beetle for a long time without disturbing it. I noticed that it seemed to collect the juice of the vine on the hairs about its mouth and then suck in the fluid. The whole process of digging away the bark and feeding on the juice was extremely interesting." In reply to questions "DR BENNETT stated that the male beetles use their cephalic horns in fighting with one another, as well as for puncturing the bark of the vines in order to bring about a flow of the sap on which they feed."

It is to be noted that the last details were not included in the report as originally written but were added from memory. It is very doubtful indeed whether it would be possible, even when "within a few feet" to detect the actual method of puncturing the bark and if we examine the broad upwardly directed ends of the head-process, which are exactly like those of the closely related *Goliathus albosignatus* shown in Plate 11, fig 3, we can hardly fail to feel sceptical as to their capability of performing this operation. The hairs about the mouth are those of the extrusible maxillae, which bear very sharp teeth, and it is much more likely that it is by the maxillae that the operation is performed. In contests between males the blunt head-processes would of course, as the most prominent part of the body, be involved but it is by no means obvious that their absence would make much difference.

64

This applies also to the mandible-horns of many Stag-beetles and other insects. The enlargement, instead of producing more effective weapons, often appears to have the opposite result. It is the jaws of the female which are strong and sharp. The great naturalist, ALFRED RUSSEL WALLACE, has described, in an interesting passage in his "Malay Archipelago" a contest between two males of the family Brenthidae with mandibles greatly enlarged in that sex. "Those curious little beetles, the Brenthidae, were very abundant in Aru. The females have a pointed rostrum with which they bore deep holes in the bark of dead trees, often burying the rostrum up to the eyes, and in these holes deposit their eggs. The males are larger and have the rostrum dilated at the end and sometimes terminating in a good-sized pair of jaws. I once saw two males fighting together; each had a fore-leg laid across the neck of the other and the rostrum bent quite in an attitude of defiance and looking most ridiculous. Another time two were fighting for a female who stood by, busy at her boring. They pushed at each other with their rostra and clawed and thumped, apparently in the greatest rage, although their coats of mail must have saved both from injury. The small one, however, soon ran away, acknowledging himself vanquished." It is the very tiny, but strong, jaws of the female which are the effective instruments and there is little reason to conclude that those of the male are of any real importance.

One last observation that must be quoted is the description by DARWIN of one of the most astonishing of all bearers of mandible-horns, the male of the Stag-beetle, *Chiasognathus Granti*, shown, together with the female, in Plate 9, figs. 1 & 2. This was seen by DARWIN in Chili during his historic voyage in the "Beagle" and he says of it "When threatened he faces round, opens his great jaws and at the same time stridulates loudly." DARWIN was evidently surprised to find that "the mandibles were not strong enough to pinch my finger so as to cause actual pain." (Descent of Man, 1901 ed., p. 461). It seems likely that, together with the utterance of loud sounds, so unusual and unexpected, the huge jaws may, in this case, have an alarming effect upon an enemy by the threat of a danger which has no actual existence and a similar effect may be produced in other cases where horns have developed to such an extent as to appear alarming. The effect, however, cannot result until the great development has been achieved and so affords no explanation of the long process by which it is attained. It must be remembered also that protection of this kind afforded to the male is of far less value to the species than would be a similar advantage to the female, who, while engaged in the vitally important task of providing for the next generation, is entirely destitute of any such protection.

These observations of the habits of the insects, incomplete as they are, reveal certain facts of importance. We have found, for instance, the extremely small number of young produced by some of them and the great expenditure of time and labour on the part of the parents required to ensure their survival. The comparative fewness of the eggs produced seems to be a characteristic of the Lamellicorn suborder, with which we are largely concerned. Instead of the hundreds or even thousands of eggs produced by many insects these may lay a dozen or even less. Their larvae are not active creatures, able like many others to seek their own food, but are entirely dependent upon the parental instinct to provide the conditions in which alone they can exist and the slightest failure in the often lengthy and complicated preparations probably entails a total failure of the next generation. The result achieved by these elaborate instincts and the bodily structures through which they operate must nevertheless be considered highly successful.

In some cases the male beetle has an important part to play in the nesting operations but this is probably not so in the majority and I have given reasons for believing that where horns have been developed to a fantastic degree there is no co-operation on the part of the male. There is no exception, however, to the rule that the existence of the next generation is absolutely dependent upon the meticulous fulfilment of the very exacting instincts of the female. This is undoubtedly the reason why in most cases the female is without the horns borne by the male, which would hinder or render impossible the performance of her functions.

As to the explanation of the various forms of beetle horns, we have been able to account for certain adaptations which are evidently useful in assisting the males of some species in the labour which falls to them, consisting of carrying materials and clearing the burrow. But it is the female who is the skilled workman and performs the delicate operations in every case and she is generally hornless. That horns are ever employed as tools there is very little evidence. Even when, as in the "Minotaur", they appear to be quite suitable, it is doubtful whether they are really employed and the fact that nearly related species of *Geotrupes* without horns perform operations almost identical with those performed by the "Minotaur" clearly shows that they are not necessary for the purpose. The implements invariably used in nidification are the legs, especially the fore-legs, and the mandibles. Where the latter are developed to such a degree as to become horns they have almost always lost the cutting and gripping power that might have rendered them useful as implements. The jaws of the male Vine-cutter, *Lethrus apterus,* have lost none of their efficiency and it seems possible that the horny outgrowth upon the lower surface may be of use when the insect is descending the vine with the nipped-off shoot gripped between them but this is uncertain and we can only admit that, up to the present, actual observation gives little or no support to the view that any beetle-horns can be properly described as tools. That a horn, the situation and direction of which are suitable, may be used in combats between rival males we have seen, but it may be accompanied by other horns which are not brought into use, and it is certain that beetles without horns often fight still more savagely than those provided with them. The extreme unsuitability of many of the horn-forms for this purpose is very evident and it is even claimed that, like the antlers of deer, they are sometimes a source of danger to both combatants. That their use, when they are used in these contests, is of any actual advantage is not very apparent nor that the hornless kinds are at a disadvantage. Finally, the complete indifference of female beetles to the rivalry of their suitors, of which the evidence is plentiful, should be noted with attention.

CHAPTER 4.

HORNS IN MALE AND FEMALE BEETLES

The information concerning the habits of horned beetles contained in the preceding chapter, if scanty, reveals certain facts of great importance for the consideration of the problem of the real significance of horns. The discovery that some horned beetles (but not horned species only) exhibit collaboration between a male and a female in nestbuilding is remarkable, for such collaboration is almost unknown elsewhere among insects. Even in completely social insects such as the ants and bees the males take no part in the work of the nest. It has been pointed out that collaboration has not been found to exist in any species of which the males are extravagantly horned and in cases in which the burrowing activities are revealed by the wearing down of the tibial teeth the difference in that respect between the two sexes indicates that such males take no part in the labour. In the Stag-beetles, the group in which an exaggerated male armament is more general and conspicuous than in any other, the genus *Sinodendron*, peculiar for the absence of the usual enlarged mandibles, is the only one in which collaboration has been found; and the Passalids, which, judging by their unusual homogeneity of bodily form, as well as their social habits, probably all practise collaboration, are exceptional in showing no tendency whatever towards the enlargement of the small horns common to both sexes.

Where participation by both sexes is found to occur there seems to be always a well-defined division of labour and the male's share is the rougher part, although it may be more strenuous. The more delicate operations are always undertaken by the female, the usually hornless sex. Although certain adaptations of the horns of males for the better performance of their special duties have been noted, these are found only in cases where the horn-development is of a very moderate kind. All the evidence seems to point to the conclusion that the Lamellicorn beetles, both horned and hornless, while they include many of which male and female share the tasks of nidification, comprise a larger number, with and without horns, of which, in common with nearly all other insects, the females alone perform all the tasks necessary for ensuring the well-being of their offspring.

Females have generally little more than slight traces of the horns possessed by the males or, in cases of enlarged mandibles, have organs much smaller though of greater strength. The only important group in which both sexes are equally horned is the Passalidae. But in other groups there are remarkable exceptions in which the female has a highly developed armament practically identical with that of the male and others still more strange in which the two sexes are quite differently armed. Of those with identical horns in male and female the genus *Phrenapates*, which so closely resembles the unrelated Passalids both in its horns and habits, is one deserving special investigation. Another is the genus *Phileurus,* also reported by DR OHAUS to live in family communities. In Australia, where about 150 different species of the great genus *Onthophagus* are found, most of them bearing horns in the male, two or three of the largest and most abundant bear an array of long pointed spikes projecting from the upper surface (e.g., *O. ferox*), and these are alike in both sexes. The great traveller and naturalist, A. R. WALLACE, in his book "Tropical Nature" suggested that the horns of beetles might perhaps serve to protect them from insect-eating birds, especially those, such as owls and goatsuckers, which hunt at night, by making it difficult for the birds to swallow them. It seems not at all improbable that the sharp spikes of these Australian beetles may be a useful protection to them from such enemies and other insects provided with similar sharp outgrowths possessed by both sexes may be protected in the same way; but the fact that it is exceptional to find any such protection in female beetles does not suggest that horns are of great importance in that way. Only those of a simple type would be suitable for the purpose, which would be defeated by any elaboration and it must be remembered that large insects are usually dismembered by birds before being eaten.

It is quite possible that, although incapable of causing actual hurt, greatly developed horns, by suggesting capabilities not really possessed, may have a deterrent effect; but again the nearly invariable restriction of such benefit to the males greatly reduces its value.

Very striking examples of well-developed horns in female beetles are supplied by two great South American insects, *Phanaeus lancifer,* Plate 5, figs. 1 & 2, and *P. ensifer.* These are splendid metallic green or blue creatures almost as big as a cricket-ball, of which both male and female have a long erect horn upon the head and a deeply hollowed thorax with a large blunt prominence behind it. At first sight the two sexes appear indistinguishable but close examination shows a tiny pointed projection on each side of the male's thorax which is absent in the female, showing that the acquisition of this great armature was not an identical process in both. Another remarkable South American beetle which, in addition to a head-horn, has in both sexes a great forked horn projecting horizontally from the back of the thorax over a large cavity beneath it, has received the name *Pinotus diabolicus.* We can only conjecture that these extraordinary forms, acquired contrary to the general rule by both sexes, may have a deterrent effect upon potential enemies.

But inexplicable as are many of the forms assumed by these strange appendages, especially those of giant insects, there are at least a few of which the meaning is clear. Many of those beetles of which it is known to be the habit of a pair to collaborate in storing provisions for their brood have the armature of the male adapted for the special function, which seems to be always his, of transporting the material dug out in the process of excavating the nest. Such habits are found in the two sub-families Geotrupinae and Coprinae, in both of which many of the males, instead of pointed horns rising from the back, have a broad hump so shaped as to produce a flat, nearly horizontal surface when the insect is ascending a rising shaft. Processes springing from the two sides may be flattened in the same plane, the pointed tips of these generally projecting outward slightly beyond the body-line, so as to form grappling-hooks for assistance when climbing with a burden upon the back. This is the structure of the male of the British *Copris lunaris,* Plate 5, figs. 5 & 6, the most accomplished of these nest-builders in our own country. It occurs also in many species of the Geotrupine genus *Bolboceras* in different parts of the world, the European *B. gallicum,* the North American, *B. ferrugi-*

neum, Plate 7, fig. 7, and others. A similar structure can be found in the subfamily Dynastinae, the male of the Malayan *Clyster Itys,* Plate 7, figs, 5 & 6, of whose habits we know nothing more than we can infer from its shape, being quite evidently adapted for the same mode of life. All these burrowing insects are very smooth and shining and are seen to be very liable to slips and tumbles in the course of their operations. Some, especially in the genus *Bolboceras,* are almost globular in shape and the value to them, in climbing a tube the diameter of their own bodies, of the little sharp projections so commonly seen is apparent. In the male British *Copris* there are two such prominences, in other species and also in various kinds of *Bolboceras* there are four and in some of the latter they are found in both sexes.

We have learnt that certain narrow-bodied beetles, Ambrosia-beetles and others that tunnel into wood, dispose of the material excavated by passing it behind the body and pushing it out backwards from the mouth of the burrow. It is found that the hinder part of the body in these insects is abruptly flattened or hollowed and generally more or less scoop-like to facilitate this. In beetles of less slender form whose burrows are so constructed as to enable them to turn round, like the horned Stag-beetle,. *Sinodendron,* the posterior end is rounded, but the anterior part, generally in the male only, is hollowed in much the same way as the posterior part of the more slender tunnellers and so serves the same purpose. This structure can be more or less clearly seen in different cases to have been arrived at by adaptation of the horns. The ridge limiting the hollowed part of the thorax of *Sinodendron* behind appears to be derived from a row of three thoracic horns, while the head, with its single hair-fringed horn, forms a movable lip to the scoop. Rudiments of the four horns can be traced in the other sex. While the female is occupied in extending her burrow and preparing the brood-niches, the male travels backward and forward, sweeping up and removing the accumulating debris with his rake-like legs and pushing it out with his scoop.

In beetles that burrow in the ground the labour involved in raising the excavated earth to the surface is much greater than that needed for the removal of wood-dust from a tree or log. Some of them are enabled by an almost globular shape to turn round easily and others which dig a large subterranean chamber are thereby spared the necessity of backward progression when returning to the surface, but all seem to raise the dug-out material by carrying it upon the fore-part of the body. In the cave-making British *Copris* (*C. lunaris*) the male has three curiously modified thoracic horns, the sharp-pointed outer ones spreading out at the base and the middle one very massive and flat in front, the effect being to form a level surface which, with the broad flat head, almost fills the round shaft in the act of ascending it. Collecting with his spade-like head the earth loosened by the female, the insect can push it before him and slowly raise it to the surface upon this convenient hod, the points of the lateral horns gripping the walls to prevent slips if necessary. Variations of this structure in many stages are found in different species, showing how adaptable are the horns.

The male of the British *Copris* has a fairly long erect horn springing from the middle of its head which seems quite unsuited to play any part in the nest-making operations. In the Spanish *Copris* (*C. hispanus*) the horn is longer and apparently more cumbersome but other forms related to these show how this seeming encumbrance has become changed in its shape and direction so as to help, or at least not to hinder, the operations. Three species shown on Plate 13 illustrate the changes. In the very common Asiatic *Catharsius molossus,* fig. 1, the head-horn is short and points forward instead of upward. In the African *C. Bradshawi,* fig. 5, the process has advanced further, the horn having acquired a rather spade-like shape, while the lateral horns of the thorax have taken a direction which seems more convenient for climbing a vertical shaft. In the Indian *C. platypus,* fig. 6, the horn has shortened and broadened still more and become a trowel-like extension of the broad flat head.

Information much more precise and detailed than it has yet been possible to obtain of the actual proceedings of any of these insects would, we may suppose, throw some light upon those very interesting cases in which both male and female carry horns and especially those in which the horns are of quite different form. The male of the Brazilian *Pinotus Mormon,* Plate 5, figs. 3 & 4, is rather similar in form to those of the species of *Catharsius* just mentioned but the modification of the horns seems to have progressed still further. The thorax is flattened out so that little

72

trace of horns remains and the head-horn is only a short pointed continuation of the trowel-head; but the head of the female bears a horn which stands straight up, as probably that of the male did in ages past, and her thorax has also a stout horn which projects forward. Astonishingly different as the two sexes are we can only suppose that both forms had a common origin, that of the female being more primitive, while that of the male has become better adapted to his special share in the work of nidification. The distantly related but much smaller *Onthophagus sagittarius,* Plate 5, figs 13 & 14, which has already been mentioned, shows a still greater disparity between the two sexes. The female has a single erect horn upon the head and the male two very small ones, while the thorax, level in front in the male, bears another stout horn in the female. Although the habits, both of this insect and *P. Mormon* are unrecorded, the worn teeth commonly seen upon the fore-legs in both sexes show unmistakably that they are hard workers and we may conclude that they work in couples, the broad backs of the males enabling them to act as carriers.

Another beetle, *Copris draco,* with horns of totally different pattern in male and female, is an inhabitant of Angola. The male bears upon the head a great curved horn, curiously toothed behind, and the female a massive crescent with its points directed upward. By comparison with the male and female *Copris* of Britain it may be seen that these remarkable appendages represent the corresponding structures carried by them, although greatly exaggerated. The male *C. draco* has also three thoracic processes like those of the male of the British species but grotesquely enlarged and obviously unfitting him for such operations as *C. lunaris* performs. It is not surprising therefore to find that the legs of the female alone show the effects of toil. It will be shown later that the possession in a greatly exaggerated form by a large insect of the same armament as a smaller related one can often be observed.

In various species of the great genus *Onthophagus,* including a little insect, *O. vacca,* not uncommon in Southern England, the females have short horns or ridges upon the head and a slight projection at the front of the thorax, while the thorax of the males is triangularly hollowed in front and the head bears a flat triangular horn, which, when the head and thorax are brought into the same plane, fits into the cavity and holds them firmly together, so forming a single broad base admirably contrived to support the load of excavated earth which, in spite of M. FABRE'S denial of such collaboration in the genus *Onthophagus,* it may with confidence be predicted it will some day be discovered that the female beetle digs out and the male raises to the surface of the ground. *)

An exactly similar modification of a short head-horn for forming a flat surface by the combination of head and thorax is shown by various other burrowing beetles of this group in different parts of the world, such as *Onthophagus rubricollis* in Australia, *Pinotus nutans* (Plate 13, fig. 3) in South America and *Catharsius melancholicus* (Plate 7, figs 13 & 14) in Africa.

But the best examples of this adaptation are found in the beautiful genus *Phalops,* which consists chiefly of African insects, many of which display lovely colours, blue, green, bronze or fiery crimson. The males have flat heads with a pair of horns but these fit so closely to the thorax that it is often difficult to see them. In the West African *Phalops Batesi* the thorax is humped and the two horns diverge so that they lie on each side of the hump; but most of the species have a triangular cavity in the thorax and the horns exactly fit this, lying side by side like the two halves of a pen-nib. The cleft between them is generally only visible on very close inspection but sometimes the two tips diverge slightly and curve gracefully upwards.

The same effect, of firmly locking together the head and thorax, is contrived in an entirely different way in some of the horned Stag-beetles of the genus *Nigidius.* In these the two front angles of the thorax appear as if deeply cut into and a narrow extension of the back of the head on each side can be inserted into or withdrawn from the two notches so formed. Like the horned mandibles, this contrivance is found in both sexes and evidently gives increased power for burrowing into the logs or tree-stumps in which they work. It is of particular interest to find that exactly the same contrivance has been acquired by very different beetles

---

*) Since writing the above I have learnt from Mr. Edgar Syms of Wanstead that he has kept in captivity specimens of *Onthophagus vacca* taken in Epping Forest and found that male and female co-operate in digging a shaft about six inches in length and leading to a horizontal food-chamber.

74

belonging to the genus *Synapsis,* which is related to *Copris.* Like the members of that genus these are mighty excavators of the earth.

The beautiful colours found in *Phalops* and many other Scavenger-beetles also serve a practical purpose. Some of these beetles, such as *Copris* and *Synapsis,* are nocturnal in their habits and these are black or dull-coloured. The bright coloured ones, on the contrary, display themselves in full daylight and several of them, including some species of *Phalops,* have been found to have a particularly offensive flavour, as shown by the behaviour of monkeys to which they have been offered as food. Amongst these gaily-coloured conspicuous forms are various species of the genus *Gymnopleurus,* which, like the Sacred *Scarabaeus* of Egypt, have the remarkable habit of pushing along the ground a ball of dung, intended as food for themselves or their progeny. This proceeding renders them so plainly to be seen that there can be little doubt that, were they good to eat, they would be quickly exterminated. Their nastiness must be learnt by experience, however, and it is therefore to their advantage to be as conspicuous as possible, so that any monkey, bird or other creature, having once tasted, will recognize and avoid them in future. One of these ball-rolling beetles in Africa (*Gymnopleurus virens*) of a vivid green or sometimes blue or crimson colour, which is found in large numbers throughout a great part of the continent, proved unmistakably nauseous to a baboon to which it was offered as a test by Sir GUY MARSHALL. Many specimens, found together in Rhodesia, proved on close examination to consist not only of this insect, a member of a hornless group, but of two others of different genera and hornless in the female alone, one of them a *Phalops* and the other the Reindeer-beetle (*Onthophagus rangifer,* Plate 5, figs 9 & 10), with its astonishingly antlered male. All three are closely alike in size, colour and shape and it is easy to understand that by mingling together in a single crowd the casualty risk to each from inexperienced tasters is reduced in proportion to the size of the crowd. The resemblance is not merely accidental, for in the Sudan is found a still more vividly coloured *Gymnopleurus* (*G. thoracicus*), of which the anterior half is a fiery crimson and the posterior half bright blue, and in company with this is found another *Phalops* (*P. princeps*), with the same remarkable coloration and of the same size and shape, while yet

75

another (*G. malleolus*), of much larger size and dull green in colour, lives in Rhodesia in company with another *Onthophagus* (*O. gibbiramus,* Plate 5, figs. 7 & 8) of the same size, colour and superficial appearance, the male of which carries antlers still more fantastic than those of the Reindeer-beetle. Although these associating pairs of insects have no close relationship their superficial resemblance is so great as to deceive at first even expert eyes. This type of mimicry, advantageous to model and mimic alike, is known as Batesian mimicry, its explanation having been first suggested by H. W. BATES, author of "A Naturalist on the Amazons". The resplendent colours, which are chiefly due to the refractive effect upon light of the microscopic structure of the surface of the body, are in some of these beetles suppressed upon the upper surface but remain in full splendour beneath, where they are invisible under ordinary conditions. We may conclude that in these cases the brilliance has proved harmful in its results and that, by the survival of the less conspicuous individuals only, continued through many generations, a less conspicuous race has been evolved. When, on the contrary, such vivid colours remain unchecked there can be no doubt that it is because conspicuousness is advantageous to the insect.

When the two sexes of a species have a dissimilar armament it may be difficult to realize how both forms can have had a common origin but it need not be doubted that this is the case and not infrequently we can trace the stages by which it has happened. There is a small beetle, *Liatongus vertagus,* not uncommon in India and China, the two sexes of which seem almost to have changed places with those of *Onthophagus sagittarius* (Plate 5, figs. 13 & 14). The male, obviously unsuited for burrowing, has a long horn, forked at the end, on his head and another, also forked at the tip and directed forward, on his back. The female, like the male of *O. sagittarius,* has two tiny horns upon her head and a high steep-fronted thorax, showing a vestige of a horizontal horn in the middle, evidently contrived for convenience of weight-lifting, like that of the male of the other species. Male and female *L. vertagus* are completely different in appearance but, if many examples are examined, small specimens of the male can be found in which, by a reversion to a more primitive condition, the long forked head-horn is reduced to a

76

mere ridge produced at each end, from which, by a slight further modification, the two horns of the female could be easily derived, while the thorax has only a very feeble anterior prominence, presenting no very considerable difference from that of the female.

We have dealt principally so far with insects bearing horns of moderate development, many of which show adaptations which are plainly in the nature of improvements for their special manner of life. Only occasional reference has been made to those extraordinary forms which have no appearance of being adapted to any purpose whatever and whose seemingly meaningless extravagance invariably causes astonishment. The adequate explanation of these and similar fantastic forms in other groups of animals has long been among the most debated and difficult problems in the whole field of Natural History. With few exceptions these fantastic forms are found in males alone but sometimes the females are horned, although less grotesquely than the males. A common characteristic of the horns of such males is their lack of uniformity. Those of small individuals often differ greatly from those of large ones and not infrequently large ones show two quite different phases. The two phases of mandible-horns in various species of *Calcodes* and *Dorcus* are shown on Plate 15, *C. aeratus*, figs. 8 & 9, for example. The same thing is found in the outgrowth-horns of other beetles. The Tropical American *Enema Pan* is a good example. Both male and female carry a strong horn upon the head, but in one male phase there is a forked horn upon the back and a pointed one upon the head and in the second an undivided one upon the back and a divided one upon the head. Such remarkable inconstancy of form seems to preclude the idea of adaptation for the performance of any special function.

Another strange instance of dissimilarity between the two sexes is the Mexican *Liatongus monstrosus*, Plate 6, figs 1 & 2, a squat toad-like insect, of which the female has two short erect horns upon the head and three upon the thorax, one in front and one on each side of a slight depression of the surface. In the male the head has only one little horn and the thorax has none but it is extravagantly wide and raised on each side into perpendicular walls, which are crowned by crests of short stiff hairs. Well-developed males and females are totally unlike each other but small specimens reveal the common origin of these two unlike forms.

77

In a small male the thorax is not wider than that of the female and the high walls of the thorax are represented only by strong ridges bordering a depression and in the same situation as the lateral horns of the female. The broadening of the thorax and enlargement of its cavity have accompanied the separation and spreading out of these horns to form the walls. The female may therefore be regarded as retaining the ancestral form of both. These beetles are found in the nests of ants. Belonging to the harmless scavenger group, Coprinae, we may assume that they live there upon friendly terms with their hosts but for what purpose is unknown. We can only conjecture that there is some connection between their structural peculiarities and their mode of life.

A deeply hollowed thorax is a very frequent feature of the male, not only in the Coprinae but in other groups. For example, in a number of very curious Australian Dynastinae the males are not unlike that of the insect just described. One of them, *Pseudoryctes dispar,* Plate 6, figs. 3 & 4, is like it, black, of similar size (about an inch long) and has the thorax extremely wide, deeply hollowed in its whole extent, and the sides greatly raised and drawn out at the summit into sharp-pointed horns curving inward. There is also a horn at the front edge of the hollowed thorax, and this, in some of the related species, assumes very curious shapes. The female, instead of the very broad hollowed thorax, has it very small and convex, and there is not the least trace of horns. But females are very rare and those of many species have never been seen. It is evident from the very powerful legs that both male and female are great diggers. Probably the females generally remain underground and the males seek them out there, for they have specially developed antennae, which are the organs of the sense of smell. What purpose is served by the hollowed-out thorax may perhaps be discovered at some future time.

The habits of an interesting beetle found in Corsica may perhaps throw a little light upon those of the Australian insects. This is called *Pachypus cornutus,* referring to the strong legs and the short horn with which the male is provided. It also has antennae with a very exceptional increase of the sensory surface. The female is a degenerate creature, with a soft body, destitute of wings and wingcovers alike and only feebly developed sense-organs. She appears to pass her time at a considerable depth below

78

the surface of the ground. The best account we possess about these beetles was recorded in 1874 by EDOUARD PERRIS (*Pet. Nouv. Ent.,* p. 383). An entomologist happening to notice in Corsica that numerous males of the species were all flying in the same direction, followed them, in order, if possible, to discover their object. He was led by them to a steep bank of hard sand into which he found they were eagerly forcing their way. Following up some of them, he discovered, at a depth of six inches from the surface, a soft-bodied wingless female. When handled, this squirted out a milky fluid, some of which fell upon the sleeve of his coat. For three days afterwards he found this sleeve attracted swarms of males and, experimenting on a later occasion by taking a female to various places where the insects were not normally to be found, he discovered that males were invariably attracted to her, apparently from some considerable distance. The potent fluid secreted by her was evidently able to bring them from afar and even when she was buried beneath the soil enabled them accurately to locate her hiding-place.

Lastly must be mentioned the rare cases in which the female bears a horn but the male is hornless. Not more than half a dozen such cases are known and these are accompanied by such peculiar circumstances as to emphasize their exceptional nature. In a few little Scavenger-beetles belonging to the genus *Oniticellus* the males have a short stout horn standing erect upon the head, but most are without this and have only a low ridge in its place. In the two species, *pallipes* and *nitidicollis,* both sexes have a strong ridge and this is a little elevated in the middle in the female, which has therefore a rudimentary horn absent in the male. In the male, however, an additional ridge appears at the back of the head and its front part is curiously swollen, as though the material which elsewhere forms a horn has here taken another shape. In the nearly related *O. africanus* the head of the male is similar and the female has only the feeblest swelling to represent the absent horn. It seems that these insects show various stages in the disappearance of a horn formerly carried by both sexes.

The case of another related beetle, found in Eastern Europe, *Onitis* (or *Chironitis*) *furcifer,* the males of which, instead of horns on the upper surface, have curious outgrowths of no apparent use on the fore-legs and the lower surface of the body, was discussed by DARWIN (Descent of Man, 1901 ed. p. 456). He says "Although the males have not even a trace of a horn on the upper surface of the body yet the females plainly exhibit a rudiment of a single horn on the head and of a crest on the thorax. That the slight thoracic crest in the female is a rudiment of a projection proper to the male, though entirely absent in the male of this particular species, is clear, for the female of *Bubas bison* (a genus which comes next to *Onitis*) has a similar slight crest on the thorax and the male bears a great projection in the same situation. So again there can hardly be a doubt that the little point on the head of the female *Onitis furcifer,* as well as on the head of the females of two or three allied species, is a rudimentary representative of the cephalic horn which is common to the males of so many Lamellicorn beetles.... We may reasonably suspect that the males originally bore horns and transferred them to the females in a rudimentary condition, as in so many other Lamellicorns. Why the males subsequently lost their horns we know not, but this may have been caused through the principle of compensation, owing to the development of the large horns and projections on the lower surface, and, as these are confined to the males, the rudiments of the upper horns on the females would not have been thus obliterated." Examples I have mentioned elsewhere of this principle of compensation leave no doubt that DARWIN'S explanation is the true one.

The most striking cases of horns confined to the female are two African species of *Onitis* (*O. Castelnaui,* Plate 8, figs. 5 & 6, and *O. tridens*), the first living in West and the second in South Africa. In these the head of the female bears a broad massive extension of its hinder edge, 5-pointed, a little hollowed behind and fitting a corresponding hollow in the front of the thorax. No doubt it serves to give greater compactness and increase the power of the head in the operation of pushing out material excavated from the burrow, as in many males of other species. *O. Castelnaui* and *tridens* are the two largest species of *Onitis* and, for that reason, the most likely to have acquired horns in the male. The males, however, retain only a rudiment of the cephalic horn but have a highly developed appendage upon the lower surface of the body. This is a great horny crescent or double-ended process which extends backward from between the front

80

pair of legs and serves no known purpose. The habits of these insects are unfortunately unknown but it seems likely that their investigation might throw light upon these very exceptional features. Probably both sexes were formerly horned but, corresponding with some change of habits, perhaps the abandonment by the male of his share in the business of nidification, while the horn of the female became better adapted for the work of transport usually performed by the male, that of the male became gradually reduced, and another outgrowth upon the lower surface appeared in its place, the result, as DARWIN suggested in a similar case, of a principle of compensation.

Of the whole vast multitude of known beetles these seem to be the only ones of which the females alone have horns. This phenomenon then is very rare and the male, in the few known cases is compensated by an outgrowth elsewhere.

It has been shown that both sexes may possess horns but to find them identical in both is rather unusual. In some remarkable cases the two sexes have each a form of its own but by far the most common is for the male alone to be horned, the female showing no more than a slight vestige. Where we have found any adaptation for the performance of particular tasks this consists in a departure from the primitive horn-form and wherever we meet with grotesque and exaggerated forms no adaptation for any conceivable purpose can be found. The variety is endless, the horns may be very short or very long, very stout or very slender, may project forward or outward, stand erect or trail behind; a few we know are used in combat but most are obviously quite unsuited for such a purpose and some can hardly be other than impediments to any kind of activity.

It remains, by a study of some of the structures themselves, to try to discover something about their origin and significance.

CHAPTER 5.

MANDIBLE-HORNS

Horns, such as those of the Stag-beetles, which are produced by a great elongation of the jaws, are comparable rather with the abnormally elongated fore-legs of the Harlequin-beetle and other insects or the produced wing-covers of male Brenthidae than with the fixed outgrowth-horns of *Chalcosoma, Dynastes* or *Sinodendron,* but when studied in detail they are found to show the same peculiar phenomena and to be subject to the same laws of growth as the outgrowth-horns. The problem they present is that of the secondary sexual characters in all their variety.

Mandible-horns are found in various different groups of beetles, but especially in the males of some, of which the females have external jaws adapted for tearing wood-fibres. Horns of this kind occur in some of the largest of the Longicorn beetles the grubs of which tunnel in wood, such as the huge *Macrodontia cervicornis,* Plate 3, the jaws of which, about half an inch long in the female, may measure over two inches in the male, the whole animal being as much as seven inches from end to end. One of the largest of existing insects, it is a creature of alarming aspect. The jaws are armed with saw-teeth along their inner edges and cross one another at the tips, which are sharp and curved like the beak of a bird of prey, while the sides of the thorax bristle with long and extremely sharp spines. But there is no reason to suppose it other than entirely harmless. Dwellers in the part of South America it inhabits charge it with nothing worse than sawing off branches of trees with its jaws and even that charge is unjustified. The saw-like appearance of the jaws has, not unnaturally, brought suspicion upon it but the cut-off branches are really the result of the operations of a much smaller beetle, which gnaws a circular groove round them and so causes them to snap off.

The *Macrodontia* is a giant compared with many small mammals, birds and reptiles but not every specimen is gigantic for some males are less than half the size of others and only a small fraction of their actual bulk.

The huge difference in size with which we are familiar in dogs, where it has been brought about by human interference, but which does not occur amongst higher animals in a state of nature, is quite common in beetles, although especially marked in certain groups in which the mode of life seems to call for no special standardization in this respect. But, surprising as the variability in their total size often appears, it is far exceeded by that of their horns. The jaws of a small *M. cervicornis* may measure one-fifth of the length of the body but those of a larger individual will form a quarter of its total length and those of the largest reach two-fifths of its length. This progressive increase in the relative size of the horns is a constant feature of great significance found in horns in general. Since the burden to be borne grows ever greater in proportion as the size of the insects is greater (not that of the individual beetle, of course, for that never grows at all, but that of different representatives of the species) it is evident that there must be a limit beyond which this process could not continue. The horns would become so cumbersome that flight first and ultimately all locomotion would be impossible.

It is usual in insects for females to be larger than their males but in *Macrodontia,* as in many other horned beetles, the reverse is found, the males reaching a much larger size than the females. The jaws of small males differ little from those of females and bear a similar proportion to the size of the body and, since that proportion increases with increased size, the striking difference between the two sexes is accounted for by the gigantic size of the male. Since with insects in general the female is commonly the worker and the male an idler, it is probable that the relatively small jaws of the female *M. cervicornis* are employed in the operations entailed in providing for her brood but it is not necessary to assume that those of the male have any use at all. They may be, like the tail of the peacock, what we are accustomed to call "merely ornamental" and it is to be noted that the peacock, like so many animals, vertebrate and invertebrate, the males of which are similarly distinguished from their females, from the beetle to the elephant, is a giant also in its group. Compared with the jaws of the female *M. cervicornis* those of the male vary, not only in size, but even more in shape and this is so to a still greater degree in some related forms. In the Indian *Priotyrannus mordax* some males have toothed and scissor-like mandibles like those of the female while others have them calliper-shaped and without teeth. Such curious inconstancy, to be found again in other horned giants, does not seem consistent with use for any very definite purpose.

An invariable accompaniment of these great horns is a greater development of the legs. Such an extension of the body in front pushes forward the centre of gravity and alters the insect's balance, so that an adjustment of its supports becomes necessary. We find therefore that, in order to redistribute the weight, the legs of the male are usually longer than those of the female, especially the front legs, which are generally longer than the four hinder ones. As a part of the distinctively male exterior the elongated legs are liable to extravagances of the same kind as those to be found in the horns and very fantastic forms may result, as in the Stag-beetle *Chiasognathus,* Plate 9, fig. 1, and the Rose-beetle *Mecynorhina,* Plate 11, fig. 4. But lengthened jaws in the male do not always entail longer legs. In the strange Indian *Didrepanephorus bifalcifer,* Plate 14, fig. 10, the mandibles, although extremely long, curve backwards over the head like the horns of a goat and its legs are very short. *Fruhstorferia 6-maculata,* Plate 8, figs. 7-9, also short-legged, has mandibles which extend forward and are very long but so thin as to be almost thread-like. These two curious insects, in spite of their very different aspect, are closely related and of similar habits, feeding upon rotting wood, like the Stag-beetles. The jaws of the females are strong and sharp but so small as to be visible only on close inspection. The varied forms assumed by those of the males show clearly that they cannot, like those of the females, serve any common purpose.

These beetles are small compared with the giant Longicorns but mandibular horns are found amongst very much smaller forms still. It is almost always noticeable however that the insects provided with them are comparatively large for the group to which they belong. For example the very abundant little insects forming the family called the Nitidulidae are familiar to most people from the very tiny beetles often found in numbers amongst the stamens of wild flowers. The family contains an enormous number of such minute insects, but a few reach a length approaching ¼ inch and amongst these are some (e.g. the genus *Psilotus*)

84

with mandibular horns in the male. Another similar group containing vast numbers of minute beetles is known as the Cucujidae. Among these are a few of comparatively large size, including the Central American genus *Palaestes,* the males of which have long slender mandibles which, as though inconveniently long, are doubled back in such a way that their tips cannot be widely separated, rendering them useless for biting or gripping. A large number of the tiny Cucujidae form the genus *Laemophloeus* and amongst these an insect little more than ⅛ in. long is a giant and may have long-jawed males. One slightly more than that length, *Deinophloeus ducalis,* has not only elongated mandibles but a pair of sharp-pointed horns projecting above them from the front of the head. Such accessory outgrowths, also confined to the male, often accompany the mandibular horns. A pair of curving horns projecting from the head in *Siagonium* are curiously like the slender curving jaws just beneath them. The drooping tusk-like mandibles of the Longicorn *Dorysthenes* almost meet a spike which projects upward from its breast, and in a species of the Rutelid genus *Dicaulocephalus (D. Fruhstorferi)* a short horn projecting in front of each eye seems as if designed to meet the corresponding backwardly hooked mandible and form with it a kind of pincer.

But it is in the Stag-beetles that mandible-horns flourish most exuberantly. In these alone they are to be found in every stage from the simplest to the most fantastic. In all other groups they are of occasional occurrence only, but in the Stag-beetles they are almost general. The group contains many kinds the males of which attain gigantic size and it is amongst these that we find horns grotesquely disproportioned even to the great bulk of their bearers. In the very small insects just mentioned the size of the male mandibles varies with that of the individual bearing them but the range of variation in the different individuals is not very great. Since it is the rule that the rate of increase of the horns is greater than that of the body, if one of these little horned forms began, from any cause, to grow a little larger in each generation the disproportionate growth of the horns would produce a more and more grotesque effect. This is the process that has taken place in the Stag-beetles and the remarkable range of variation in size prevailing in many of them shows us various stages in their past history.

There are certain small forms of Stag-beetles (Lucanidae), especially the two genera *Figulus* and *Nigidius,* in which the mandibles are alike in both sexes. In the latter genus they bear branched or hooked appendages. But in the great majority there is a very marked sexual dimorphism — the females have small but strong jaws and the males much longer but less powerful ones. The passage from the female to the male form of mandible is not as complete as in the horned Longicorns. Usually those of the male, even in small specimens in which they are scarcely longer than those of the female, are rather differently shaped, less solid and meet less closely. In many kinds, like the great British *Lucanus cervus,* Plate 8, figs. 1 & 2, there is no resemblance between them although it cannot be doubted that both sexes of their remote ancestors were alike. But in a few of the small members of the family we are able to see the process of differentiation at its beginning. In the island of Mauritius there is a very rare little species, *Vinsonella caeca,* less than half an inch long, which is found only upon the top of a mountain crest. There is a very slight difference between male and female in the shape of the head but the mandibles are alike except that those of the females are straight and those of the males are a little longer and curve gently upward. The strange thing about this little insect is that it is totally blind and the wings which enabled its ancestors to fly have degenerated to a condition in which they are quite useless. In its lofty island habitat the power of flight must have meant the risk of being blown out to sea and only those that abandoned the habit of flying have survived. Perhaps blindness also acts as a check upon any disposition to stray from the narrow confines of their mountain-top, but it is rather remarkable that, in spite of such seemingly severe disabilities, the members of the community can still retain contact with each other. In South Africa are also found curious Stag-beetles (*Colophon*), of larger size and with strangely horned males, each species of which is similarly confined to a single mountain summit. These also are unable to fly and, although not sightless, their eyes are very small and probably of little use.

The strange inconstancy of shape in the mandible-horns of some of the Longicorn beetles has been mentioned but that occurring in the Stag-beetles is far greater. In these also it depends upon the size of the individual, which in the males is remarkably variable. Females of the British *Lucanus cervus* measure from an inch to about 1½ inches in length, but a large male may be as much as 3½ inches long, half its total length being occupied by the head and horns, while a small male may be no more than 1¼ inch long, that is little more than one third of the length of a large one and perhaps not more than one-twelfth of its weight, with jaws about one-fifth as long. Of some foreign species (e.g. *Dorcus arfakianus* in New Guinea, Plate 14, figs. 1-7) small males are much less than one-third of the length of large ones and their mandibles less than one-sixth, while in the females variation in size is small and in the mandibles not appreciable. The difference between the two sexes is sometimes so great that they are practically unlike in every respect, so that there may seem to be no reason at all for associating them. The shape, the colour, the character of the surface of the body, may all be different. For example the West African *Homoderus Mellyi,* Plate 14, figs. 11-13, bright yellow in the male, except for four black spots on the thorax, has a black-headed female with black wing-covers, each decorated with a yellow stripe, the head, broad and smooth in the male, is narrow and very rough in the female and the legs and antennae, as well as the jaws, are quite different.

A striking discrepancy usually found between the two sexes of Stag-beetles and other mandible-horned insects is in the size of the head, so astonishing in the common British Stag-beetle, *Lucanus cervus,* and an evident consequence of the development of the horns. With these it varies according to the size of the specimen. In the small *Homoderus Mellyi* (fig. 12) the head, although larger than that of the female, is the narrowest part of the body, but in a large specimen (fig. 13) not only has it become the widest part but it bears a great crest twice as large as the whole head of the small example. The mandibles are also differently shaped so that, unlike as are male and female, small and large males are no less unlike. As a consequence of these remarkable differences between different forms of the same species, when, in old days, only a few specimens of any species were usually known, such unlike examples were inevitably regarded as belonging to different kinds of insects and given different names. It is only when a considerable series has been brought together that the various male phases are seen to form a regular progression and to accompany differences in size.

Small males often bear greater resemblance to the female form than to large males of the species, as seen in *Dorcus Reichei* (Plate 15, figs. 12-19). The female of this (fig. 12) is smooth and glossy in front, except upon the head, and corrugated and rough behind. The male, if of large size, is smooth and dull, with long mandibles (fig. 19), but a small male (fig. 13) is like the female, except that its mandibles are thin. Successively larger males show a gradual change from the female aspect to that of the large male, although small and large seem to have little in common.

The range of size in the males of these beetles is remarkable but the varying size of their horns is much greater and most surprising of all is the difference in pattern to which the horns are subject. Plate 15, figs. 1-4, show four males of a Ceylon species, *Calcodes carinatus,* each, it will be seen, with mandibles unlike those of the others. But in this series another curious circumstance is brought to light. Between the first three specimens there is a regular progression in size and in a long series the gradual transformation of the mandibles can be traced; but figs. 3 & 4 show specimens of the same size, the largest size to which this species attains, and between the two forms of mandible no intermediate is found. Why this strange break of continuity occurs is unknown but it is no rare phenomenon. It is found in many Stag-beetles, as well as in horned beetles of other kinds. Figs. 5 & 6 show two males of *Calcodes Siva*, a species related to the last, inhabiting Assam. Both specimens again are of the full size attained by the species and their mandibles are very different but no connecting links are ever found. The most striking case of this strange dimorphism of the male is perhaps that of *Calcodes aeratus,* (figs. 7, 8 & 9), an insect found in the Malay Peninsula. The specimens again are all males and the second and third are of full size. A slight elongation of the mandibles is visible in the second specimen. Although very slight, it shows the full extent of variation accompanying the increase in size of this species. The remarkable isolated phase shown in fig. 9 appears abruptly and is totally different. No connecting link between it and the ordinary form ever occurs.

So complete is the gap that separates these two phases of the large

88

males that to regard them as distinct species as formerly was perfectly natural. That they are not so is conclusively proved by certain rare examples in which both phases appear upon the opposite sides of a single insect. Two cases of this are shown in Plate 15, figs. 20-25. The first, *Dorcus forceps,* inhabits Borneo and Sumatra. Fig. 22 shows a male of the ordinary variable phase of this and fig. 20 the isolated or constant phase, while between them appears a specimen of which the right mandible is that of the isolated phase and the left that of the variable phase. Of the next species, *Dorcus polymorphus,* a beetle abundant in the Darjeeling district of India, of 80 males sent to me for examination 77, representing all sizes, had the triangular mandibles seen in fig. 25, the inner edges meeting throughout their length in small examples but a little separated at the base in large ones. Of the other three, two had mandibles as in fig. 23, slender, curved and meeting only at the tips, with a strong erect tooth at the base, of which the normal form has no trace. The last specimen, which is shown in fig. 24, had on the left side the curved mandible of the isolated phase and on the right the triangular mandible of the normal phase.

One of the most beautiful of the many beautiful beetles to be found in Australia affords another instance of this kind. This is the gorgeous *Lamprima Latreillei,* which is sometimes seen disporting itself in large numbers in hot sunshine, different individuals being of the most vivid blue, green, golden or fiery red colours. Being so conspicuous but making no effort to conceal themselves, it is probable that they are unpalatable to birds and other insectivorous creatures and their brilliant colouring serves to advertise that fact to any that have made the experiment and so to render its repetition unlikely. The mandibles of the males are much longer than those of the females, curve upward and bear a single tooth, which is not sharp but very blunt and partly concealed by a thick pad of velvety hair. In 1885 a beetle was found in North Queensland with many teeth along the edge of its mandible instead of a single one. This distinguished it as belonging to the genus *Neolamprima* and it was accordingly named *Neolamprima mandibularis.* Years afterwards it was noticed that a swarm of these beautiful creatures consisted partly of *Lamprima Latreillei* and partly of *Neolamprima mandibularis,* although

89

7

the females were all alike. It is now known that *Neolamprima* is one of the phases of the male *Lamprima* and a specimen has been found bearing one mandible of each phase. Plate 9, figs. 4-7 show *Lamprima Adolphinae*, a nearly related New Guinea species with two corresponding male phases, but the long-jawed form is the common one, while the other is very rare. The closely-set teeth of these long male mandibles much resemble those of the corresponding phase of *Calcodes aeratus* and speculation as to some important function being served by the curious apparatus is naturally aroused; but the problem is complicated by the fact that in these two beetles this phase is found only in a small minority of males and in *L. Latreillei* seems to be absent altogether in most places inhabited by the species, although in its New Guinea representative it is the normal phase and the other very rare.

Yet another difference in the male mandibles of a single species is shown in Plate 15, figs. 10 & 11. These are two forms of *Dorcus giraffa* found in different parts of the great area over which the beetle ranges in Asia, the first from the Himalayas and the second from Assam, Burma and the Malayan region. The Himalayan form, with gently curved and many-toothed mandibles, was given the name *Arrowi* and its female was reported as quite different from that of the other form with fewer and stronger mandibular teeth and abrupt curvature, but unfortunately the female described was that of quite another species and actually there is no difference whatever except in the male mandibles.

The extraordinary inconstancy of all these male insects in the size and pattern of their horns is assuredly their leading characteristic. Why such diversity should have been developed and such various forms exist in a single species is an unsolved problem. The females, although varying to some extent in size, are otherwise constant and their mandibles scarcely vary unless worn down by use. Does not this remarkable difference between the two sexes indicate that, whereas the mandibles of the females have assumed their particular form in adaptation for the work they are called upon to perform and which would be less efficiently performed by any departure from it, those of the males are under no such restraining influence and their form and size, within very wide limits, are of little importance or none?

There is another curious fact to be noted, which seems to point to the same conclusion. The mandibles of female Stag-beetles are generally very sharp-pointed and bear at the inner edge one or more stout teeth and these teeth differ a little on the two sides, so that the jaws are not symmetrical. The effect is that they can be brought closer together and grip more tightly any object held between them. The mandibles of small males often show the same absence of symmetry and this may be exaggerated in those of larger size, but very large ones, showing the highest development, are almost always exactly symmetrical. Symmetry is gained at the sacrifice of the gripping power obtained by asymmetry, which, important for the female, has ceased to be of importance to the male. The enormous elongation occurring in the male mandibles has a similar significance. The short jaws of the females have great strength but the amount of force that can be exerted at the ends of those of the male diminishes in proportion to their length and, as Darwin found in the grotesque *Chiasognathus Granti* (Plate 9, figs. 1 & 2), whose horns, like those of *Lamprima Adolphinae,* are finely toothed, becomes very inconsiderable. It is of course quite possible that some of the species, in which elongation has reached less extravagant lengths, may still retain sufficient strength in the jaws for practical purposes.

A few words may be devoted, in conclusion, to pointing out the close analogy, in the sexual differences which have been described in these mandible-horned beetles, with those of a very dissimilar and unrelated group, the curiously elongated Brenthidae. As described by A. R. WALLACE, the females (see Plate 4, figs. 2 & 11) bore holes in trees, in which they deposit their eggs, their very minute mandibles, placed at the end of the slender proboscis, operating like the point of a gimlet. The males take no part in the operation and many of them have undergone very remarkable changes of form (figs. 1 & 10). The head is grotesquely enlarged and the mandibles, which a further increase would entitle us to call horns, are enormously larger than those of the females. When compared with the sharp cutting instruments of the female, however, the change appears as a kind of degeneration.

CHAPTER 6.

OUTGROWTH-HORNS

Outgrowth horns are more varied than mandible-horns in their forms and situation. There are many groups of beetles, such as the Carabidae, Staphylinidae, Tenebrionidae, Anthicidae, Erotylidae, Cisidae, etc., of very diverse character and habits, in which horns are found exceptionally but in which they are always of a simple type. In all these cases the insects are of small or quite moderate size although very commonly larger than hornless species closely related to them. In strong contrast are those which create an immediate impression of extravagance and grotesqueness. These all belong to the same group, the Scarabaeidae, and chiefly to a single section of it, the Dynastinae. All of them are very large insects and they include the largest to be found in the world to-day. Between these two groups is a third, belonging also to the Scarabaeidae and exhibiting horns generally highly developed and very varied in form but without the extravagance of those last mentioned. These are upon the whole intermediate in size between the first and second groups and nearly all are found upon the ground, in which they excavate burrows often of remarkable depth and extent. It is amongst them that is found the surprising fact, almost unknown elsewhere in the animal kingdom except in birds, of nest-building performed by a male and a female in collaboration. Nest-building operations of astonishing complexity are of course performed by insects of many kinds but almost invariably by the females alone. To find, as we have seen is not uncommon amongst Lamellicorn beetles, a pair of insects sharing the labour is surprising indeed. A fact of particular significance is that, both when the male participates in the work and when he does not, the female is the skilled worker, whereas, with few exceptions, horns are confined to the male.

It will be noted that in the three groups above referred to there is a general correspondence between the size of the insects and the degree of development attained by their horns. In other respects there is no

92

relation whatever between the size of an insect and its developmental status but we shall find later that, in regard to horns, a correspondence of this kind exists both between related species of the same group and between different individuals of the same species.

The most important materials for the study of outgrowth-horns it is evident are to be found in the Lamellicorn beetles, the group in which the insects attain their maximum size.

The Lamellicornia or leaf-horned beetles are so-called from the fact that their feelers or antennae have the terminal joints lamellated like the leaves of a book and so capable of being separated and brought together again in a similar way. By this means the more sensitive part of these, the most important of their sensory organs, has its surface both increased and protected and perhaps, by reason of such protection, has become more delicately sensitive than the corresponding part in many other insects. This is the seat of the olfactory sense, the chief means by which insects are able to locate both the other individuals of their species and the substances that serve as food for themselves and their progeny. It is possible that this special sensitivity has some connection with the highly elaborated instincts of many of the group. Examples of its astonishing efficiency have already been given.

The Lamellicorn beetles consist of three families of which the Lucanidae and Passalidae together are far less numerous than the third family, the multitudinous Scarabaeidae, amongst which the majority of all the known bearers of horns on head and thorax are to be found. About 25,000 different kinds of Scarabaeid beetles have already been distinguished and furnished with names. Of these there are nine principal sub-families, five of which abound in horned forms, while some are found in nearly all. In two large groups, Rutelinae and Melolonthinae, they are few. These are the groups which feed in the main upon living plants and are in consequence potentially enemies of mankind. As a general rule it may be said that the horned beetles are useful insects. Many of them feed, like the Stag-beetles (Lucanidae), in dead trees or tree-stumps and, by hastening the decay of these in forest country, help to clear the ground for fresh growth. The rest are mostly scavengers, feeding upon waste and decaying substances, animal and vegetable, which, in many cases they first bury

93

beneath the ground as provision for themselves or their brood.

Although this great section of the Coleoptera contains a vast number of horned forms, the distribution of these is very irregular and it is certain that their horns have had no common origin. Some bear them upon the head alone, others upon the thorax alone and some upon both head and thorax. Whether upon head or thorax they have arisen in different ways and the prevalence of the feature in Lamellicorn beetles is not due to common descent from horned ancestors but to a special tendency to the acquisition of horns possessed by the group. This is probably connected with the fact that many of the largest insects are found in the group, since, as just mentioned, the possession of horns very often distinguishes species of comparatively large size.

That the Lamellicorns are descended from common ancestors there can be no reasonable doubt nor can it be doubted that their ancestors in the distant past were burrowing insects, for the whole group has retained a fundamental structure adapted for burrowing. Since in the grub-stage all live underground or hidden within the substance of a dead tree-stump or similar situation and have very slight power of movement from place to place, it is necessary for the mother to burrow in order to lay her eggs in the proper situation to ensure the well-being of her progeny. The burrowing habit requires a certain degree of hardness in the exterior of the body, at least in the head and fore-part, as well as in the legs, the chief digging implements. The fore-legs often have the outer edges provided with scraper-teeth but in the males these teeth may be absent or present in various degrees of degeneration. In some of those beetles which, like the Stag-beetles, feed upon wood hard and strong mandibles are essential and these, in the males, have often developed into horns ; in others, such as the Rose-beetles (Cetoniinae), many of which eat the stamens of flowers, the jaws are only adapted for dealing with soft substances but the head is pointed or spade-like to fit it for burrowing and this part, in many males and a few females, gives rise to horns of many shapes and sizes. In some, like the African *Mecynorhina Polyphemus,* Plate 11, fig. 4, and *Ranzania petersiana,* Plate 12, fig. 6, the front margin is produced into a horn ; but the side margins may form a pair of horns, as in the beautiful Indian *Cyphonocephalus,* Plate 12, fig. 4, *Narycius,* fig. 3, and

Dicranocephalus, figs. 1 & 2. The horns may taper to a point, dilate at the end or become forked or branched. In the genus *Trigonophorus* a curious stalked leaf-shaped horn rising from the front margin is found in male and female alike but another little horn rising from the back of the head is pointed in the male and blunt in the female. In the wonderful golden Theodosias of Borneo, although it is quite unusual in this group, besides a slender horn upon the front of the head there is another projecting above it from the thorax. In *Theodosia magnifica* both horns are pointed and in *T. telifer* both are forked.

In many of the males with curiously developed heads the forelegs also are curiously modified. As in those with greatly enlarged mandibles, the extension of the head has the effect of throwing the centre of gravity of the body farther forward than in the female. This has necessitated an adjustment of its supports by the elongation of the forelegs and sometimes this has been carried to an extravagant length or accompanied by fantastic embellishments, as in the grotesque *Mecynorhina* Plate 11, fig. 4, or the gaudily decorated *Cheirolasia* Plate 12, figs. 7 & 8. It is not possible to suppose that these extravagances, either of head or forelegs, have any practical use. They seem to indicate only that these parts of the body, important in the female, have ceased to be so in the male.

In other cases the elongated legs, instead of a more elaborate form assume a simpler type than those of the female, by the disappearance of the lateral teeth always present upon the female tibia. In the great Goliath beetles of Africa, the males of which have a forked protuberance upon the head, there are always three sharp teeth upon the front tibia of the female but none in the male; but in *Goliathus albosignatus,* Plate 11, fig. 3, the smallest of the giant species, small teeth although normally absent may sometimes be seen in small-sized males. In the medium-sized *G. Fornasinii,* fig. 2, only one of the teeth is lost in the male and smaller related forms with a smaller but similar horn have teeth in both sexes, showing that these are nearest to the ancestral form from which all have descended. In the fantastic legs of *Mecynorhina* and *Cheirolasia,* just mentioned, the degree of development also depends upon the size of the specimen, so that in both horns and legs we find the same progression according to size as in the mandible-horns of other beetles.

In other horn-bearing Lamellicorns horns of similar origin occur upon the head and are often accompanied by others upon the thorax or they may be found upon the thorax alone. In many of the Rhinoceros-beetles and Scavenger-beetles the thorax of the males is deeply hollowed out and the fold of horny substance forming the margin of the hollow may be drawn out to form a horn behind, as in the extraordinary *Onthophagus imperator,* Plate 5, fig. 12, at the two sides, as in *Ceratoryctoderus,* Plate 6, fig. 7, and *Trichogomphus robustus,* fig. 8, or in front, as in *Strategus Simson,* fig. 5, the last having lateral horns as well. Yet another horn may be borne upon the head, as in *Dipelicus Geryon,* Plate 13, fig. 4, or a pair, as in *Onthophagus imperator.* In the great Hercules-beetles, Plates 2 and 1a, three such horns have been produced but the enormous enlargement of the middle one has had the effect of obliterating the cavity from the hinder margin of which it was produced. A hairy cavity, flanked by the lateral horns and continuous with the hairy and hollow lower surface of the great horn, still remains in *Dynastes Neptunus,* Plate 1a, but in *Dynastes Hercules,* Plate 2, the great horn has developed further, carrying the lateral horns with it, and the cavity has completely disappeared ; but, although inverted, it is still represented by the hollowed lower surface of the horn with its brush of hairs. That the advance of the lateral processes on to the horn is due to the drawing out of the thorax is shown by the situation of these processes at the base of the horn in small males and at intermediate distances along it in others of medium size, as well as by their intermediate position in other but closely related species.

Comparison of various species of the remarkable South American genus *Golofa,* shown in Plate 10, reveals a similar change. In *G. argentina,* fig. 6, the male has a hair-filled cavity upon the thorax, its hinder edge sharp and sometimes a little drawn out, but in most of the species this edge has extended into a long horn clothed with hair on its anterior face. In *G. claviger,* figs. 3 & 4, the cavity has vanished and the horn is erect, expanding at the end into a kind of canopy, which is hollow and hairy beneath, a very extraordinary inversion of the original hollow.

All these curious forms, in addition to the dorsal processes, have a horn upon the head, of simple form in small specimens but in large ones elongated and more or less embellished with teeth, few or many. In

Dynastes Neptunus there is a row of small teeth, in *D. Hercules* two large teeth in specimens from Mexico or South America and three in those inhabiting the West Indies. In *Golofa claviger* traces can be seen of two rows of tiny teeth. In another related species, *Golofa Porteri* Plate 10, fig. 1, in which the horn is exceedingly long and slender, the teeth are beautifully regular and the appearance is that of a double-edged tenon-saw, which has earned for the insect the name of the Sawyer. The thoracic horn of *G. Porteri* is even more slender than its head-horn and the expansion which terminates that of *G. claviger* has disappeared but the soft hair clothing the lower surface remains as a delicate fringe along the inner edge of the fragile horn.

How and why these and other strange forms have come to be are problems we have hardly begun to solve. Although in their less developed states the horns often look like instruments adapted for some aggressive purpose they frequently point upwards or backwards or in some other quite unsuitable direction and when we turn from such dwarf forms to those of full development, all appearance of adaptation to any practical use is usually lost. When, as is sometimes the case, there is an appearance of gripping power between the horns of the head and thorax, as in *Dynastes Gideon,* Plate 8, figs. 3 & 4, where both horns are forked at the end, producing four opposing points, they may actually possess the power of gripping and use it, as has been described, in some individuals; but, so variable is the degree of development, that this possibility is almost certainly confined to a minority of the specimens. In the species most nearly related to *D. Gideon* the minor development or complete absence of horns renders any such use impossible, as it is in the vast majority of horned forms, in which the horns are only borne by the head or thorax or, if by both, are of such a shape as to render gripping impossible. It is probable that the curious habit of the male *D. Gideon* of attempting to grasp the female between its two horns is a kind of play, comparable to the dancing of gnats or the chirping of crickets, which the particular form of its horns enables it to indulge in.

The deep cavity in the thorax, from the margins of which horns are so often derived, is a common feature in very different Lamellicorn groups and almost always found in males alone, although females may show it in a rudimentary form. From the proofs of its former existence in insects which have now almost lost it, it seems that is was more general in the past than at present and it is evidently more primitive than the horns that have arisen from its margins. Whether it has or had any practical use has not been discovered. The habits of the insects of the present day may differ considerably from those of their remote ancestors or the latter may be perpetuated only by a few unknown or little known survivors. Members of a South American genus of Scarabaeinae (*Eurysternus*) are remarkably flat above and have been reported by OHAUS (*Stett. Ent. Z.*, 1909, p. 72) to pile upon their backs a load of the excrement upon which they feed and, holding it there with the middle pair of legs, carry it away, walking upon the other two pairs of legs. The backs of most beetles are smooth and rounded and, if this is the explanation of the peculiarly flattened back in *Eurysternus,* the hollowed out backs of other beetles, perhaps in times past, may have also served for purposes of transport; but of this we have at present no evidence at all.

The Scarabaeid genera *Liatongus* and *Drepanocerus* contain a variety of striking cases of these hollow-backed males, curiously horned at the margins of the cavity. Most of them have a pair of lateral horns, while a third is borne upon the head, but *Liatongus Rhadamistus,* a vividly coloured orange and black South Indian insect, being without a head-horn, has the front edge of the cavity as well as the side margins drawn out. In *L. mergaceras* and others it is the hind margin that is produced and in some the continued growth of the hind marginal process has resulted in obliterating the cavity, exactly as in the great Dynastinae. The genus *Drepanocerus,* found in Africa, India and China, is remarkable for containing the smallest of all beetles with highly developed horns. All its members are tiny, no larger than a barley-corn but the males of some of them bear very curious appendages. The Chinese *D. sinicus* has the side walls of the hollow thorax raised into a pair of relatively large hook-like horns pointing backwards, *D. setosus* in India and Ceylon, has a single horn, divided at the tip, rising from the hinder edge of the cavity and pointing forwards and *D. Kirbyi* in South Africa has three forward-pointing processes, of hair-like fineness and half as long as the body, one arising from each side of the thorax and the third from the front of the

98

head. For such minute insects so well developed an armament is surprising and must have a special explanation. The beetles, when found, are invariably caked with dirt and the surface of their bodies bristles with tiny hairs or scales which are often hooked and seem to have the object of entangling and holding this, probably for concealment or to make them less attractive morsels for any insectivorous creature likely to attack them. It seems possible that the horns borne by the males may assist in this, although the males hardly need more protection than the females. Is it possible that this is another method of transporting food-material into the nest, to be removed on arrival there, as no doubt happens in the case of *Eurysternus*? The dirt-encrusted condition is certainly not without meaning, for insects in general are cleanly creatures and the Scavenger-beetles, in spite of the unpleasant materials with which they occupy them-selves, are normally found in perfectly clean and even resplendent con-dition. Like most insects, they spend much time in very carefully and systematically purifying themselves, using their mouths and feet, like cats, for that purpose.

That the horns of beetles sometimes come into play in contests between them need not be doubted, although the absence of any special adaptation for such a purpose is, in most cases, so obvious. We have even some evidence that, as in those of stags, their horns have sometimes become entangled, with unhappy results for both combatants. But that aggressive employment is very uncommon appears certain from the fact that marks of damage such as must result from it are practically non-existent in collections containing thousands of specimens and although the horns are often so fragile. They are not solid, but thin-walled and hollow and, though sometimes fairly sharp-pointed, very brittle.

The extraordinary genus *Onthophagus* of scavenging beetles, with its 1600 different forms in all parts of the world, shows in one or other of them nearly every type of horn on head or thorax known in beetles. None of these are giants among the insects and some of them are very minute but they illustrate clearly the close connection between horns and the size of their bearers for, while the smallest of them are without horns, most of those of medium size are provided with them, generally of a simple type. It is in the largest members of the genus that the most remarkably developed and fantastic forms are to be found. A few of these have been already mentioned, such as the Reindeer-beetle, *Onthophagus rangifer,* Plate 5, figs. 9 & 10, whose long slender antlers, trailing backwards almost to the end of the body, branched a little way along and knobbed at the end, are plainly not designed for practical purposes and, if involved in combat, would surely prove more hindrance than help. But others are still more fantastically armed. *Onthophagus Elgoni,* for instance, in addition to similar filamentous antlers, with the shorter branch forked and the longer one knobbed, has its thorax deeply hollowed and from each side-wall of the hollow a flat, sword-like blade, with a rounded end, rises straight upward. A more war-like aspect is presented by *O. pyramidalis* in Rhodesia, *O. imperator,* Plate 5, fig. 11, in southern India and others, in which a single sword-like blade thrusts forward from the hind margin of the cavity; but in all these insects the weapon is blunt at the end and the long antlers upon the head look as though specially devised to impede its use. There is no such impediment in the Abyssinian *O. Blanchardi,* whose head is hornless, but the thorax bears a lance-in-rest projecting straight forward, sometimes far beyond the head, when it is half as long as the body. But it has no real strength and its end is blunt. Moreover many specimens present a different appearance altogether. Amongst 39 examples in the British Museum 15 only have a long implement reaching beyond the head; in the other 24 it only reaches the head and in 9 of these it is a mere vestige. It is evident that the same purpose could not be served by both long and short instruments and, if either is really serviceable, many, if not most of the insects, are ineffectively provided. It seems an inevitable inference from this strange inconstancy that the value of the structure to the beetle cannot be great.

As a result of this inconstancy most of the horned beetles which have been longest known have received several names, the early systematists quite naturally having regarded the odd specimens which first presented themselves as representing several species. The comparison of long series of specimens, including all degrees of development, was necessary to show that differences which in other organs would have had that significance were without it in these.

Comparison of a sufficient number of individuals soon reveals that the different degrees of horn-development are closely related to differences in body-size. The variation in size is nearly always greater in males than in females but is much greater is some than in others. It should be noted that those cases of adaptation of the horns for a special purpose which have been described here are all cases of insects of which both sexes may be supposed to be burrowers and in no case such as have attained gigantic size, with the accompaniment of extravagant horn-development. For insects called upon to perform laborious tunnelling feats, like the British *Copris* and *Geotrupes Typhaeus,* the diameter of the body determines that of the burrow and even a slight increase of body-size entails a great increase in the material to be removed from the burrow, and the time and effort required to remove it. This consequence seems to have brought about a greater standardization of size in these than in others under no such restraint. It is in those great insects in which the horns are most fantastically developed that the size of individual males and, with it, the degree of development of the horns is most variable. In such forms as *Dynastes Gideon* or the Atlas-beetle (*Chalcosoma*) the horns, so immense in a large specimen, may be almost entirely wanting in a small one.

If as many males as possible of one of these great beetles, *Dynastes Gideon,* for example, are arranged in the order of their size it will be found that they are marshalled almost exactly in accordance with the degree of development of their horns ; but whilst the beetles, measured from chin to tail, vary in length from about 30 to 50 mm., the thoracic horns, measured from their tip to the base of the thorax, have a length varying from 8 to 40 mm., i.e. for less than twice the body-length the horn is five times as long. In fact, in the smallest known specimens the horn is practically absent altogether. Comparative measurements of the giant Hercules-beetle (*Dynastes Hercules*) of Central and South America, which also has horns upon both head and thorax, are still more striking. The body-length of the largest males is about 80 millimetres and the thoracic horn extends for the same distance in front of the head, making the total length of the insect about 6½ inches; but a small male with a length of 54 millimetres has a horn measuring only 6 millimetres, or only one thirteenth part of that of his bigger brother.

The species of larger size reaches a more disproportionate growth than the smaller one, but the same principle applies to all, that while the size of the horn depends upon that of the individual specimen, its proportionate increase is much greater than that of the body. It must not be overlooked that in the males of *D. Gideon* and *D. Hercules* there are two horns, and in the former, while in large examples that of the thorax is longer than that of the head, it is noticeable that with diminishing size the shrinkage is more rapid in the thoracic than the cephalic horn. In small ones the head-horn is longer than the other, remaining distinct, though small, when this has disappeared. This longer persistence points to the conclusion that the generally smaller horn was an earlier acquisition than the larger one in the past history of the insect. *Dynastes* (or *Xylotrupes*) *Gideon* was the subject of a statistical investigation by BATESON and BRINDLEY (*Proc. Zool. Soc.*, 1892, p. 589), who measured the length of the two horns separately in 342 male specimens in order to discover what proportion those of different lengths bore to the whole. The wing-covers were also measured, as a standard of the actual length of each specimen, and it was found that, as might be expected, those of average size predominated. From the horn measurements however the conclusion reached was that moderately long and moderately short horns were most numerous and those of average length comparatively few. A more reliable result would perhaps have been obtained if, instead of being taken seperately, the combined lengths of the two horns had been used, and still more if it had been possible to measure the actual bulk or weight of the horns. Their length is only a rough indication of this, since they vary not only in length but in thickness also. A specimen may sometimes be found with a longer horn than another of larger size, but it will generally be noticeable that in such a case the shorter one is stouter than the longer. The forelegs too, of *D. Gideon*, have an elongation in the male which varies in the same disproportionate manner as the horns and in any mathematical calculation this also must be taken into consideration.

It was noted by LAMEERE that, as a result of the correlation between the body-size and horn-development, when several nearly related species differ in average size they will often show a progression in their horn-development similar to that of different-sized specimens of a single species. For example, the thoracic horn of a full-sized *Dynastes Hercules* of a body

102

length of 3¼ inches measures 4 inches, that of *D. Neptunus,* a rather smaller species, also South American, reaching barely 3 inches, is 3 inches long, that of the Mexican *D. Hyllus,* 2½ inches long, is about 1½ inches and that of the North American *D. Tityus,* not quite 2 inches long, is about 1 inch. Males of *D. Hercules* of corresponding sizes show similarly differing lengths of horn. But that the mere length of the horn is not always a suitable standard of comparison is shown by another North American species, *D. Granti,* still smaller than *D. Tityus,* in which the horn is a little longer but at the same time less stout and without the lateral offshoots seen in all the other species. In the extraordinary genus *Golofa* a similar progression can be traced in the degree to which horns have developed in different species. On Plate 10 are shown five species inhabiting different parts of America, arranged according to the average size of each. The smallest, *G. inermis,* fig. 7, is quite hornless, in *G. argentina,* fig. 6, the head bears a slender horn and the thorax has a cavity filled with hair, its hinder edge rather sharp. In the rather larger West Indian *G. Guildingi,* fig. 5, this edge is drawn out into a curiously shaped lobe which bears the hair-filled hollow beneath, while in the great *G. claviger,* represented in figs. 3 & 4, the lobe has become a massive horn, still hollowed and hairy underneath, while the dorsal cavity which it represents is obliterated. In *G. Porteri,* Fig. 1, the largest species of the genus, although the horns of head and thorax are the longest of all, they are also astonishingly slender, so that their combined weight is probably less than that of the smaller horns of *G. claviger;* but in *G. Porteri* there is an extraordinary elongation of the fore-legs, peculiar to the males like the horns, and this may be regarded as compensation for the thinness of the horns. In DR JULIAN HUXLEY's book, "Problems of Relative Growth", fig. 94, outlines of four species of *Golofa* are given to illustrate this progressive horn-development but *G. imperialis* and *G. Eacus* (not *coecus* as printed) are really two species of the same average size. The horns of the former are short and stout and those of the latter very long and slender, so that the total result is about the same in each case. This kind of compensatory development may often be seen in nearly related insects and, as DARWIN concluded, explains the replacement of the horn in the now hornless male *Onitis* by an outgrowth elsewhere in the body (see *O. Castelnaui,* Plate 8, figs. 5 & 6).

The Eastern *Dynastes Gideon* is a much smaller insect than the American *D. Hercules* and *Neptunus* and, striking as its horns are, they do not bear comparison with those of these giants. There are other Eastern species smaller than *D. Gideon,* the horns of which are quite small, while in the smallest, *D. inarmatus,* found in Java, they have almost vanished. In Japan and Eastern China lives *Dynastes* (or *Allomyrina*) *dichotomus,* which is a little larger in average size than *D. Gideon.* It bears on its head an immense horn which expands at the end into a broad four-pronged fork but its dorsal horn, compared with that of *D. Gideon,* is greatly diminished (see Plate 13, fig. 7). The male of another curious beetle, *Lycomedes Reichei,* found in Colombia, has upon its head a long horn, forked at the end and bearing a sharp tooth behind, and upon its back another standing upright, hollowed in front and looking rather like a spoon balanced upon its handle. A nearly related species of similar size has the dorsal horn reduced to a mere hump but the head-horn is very stout and the tooth has become a strong branch, which itself divides again. This is called *Lycomedes ramosus.*

The male of the Asiatic *Onthophagus tragus* normally bears two horns upon the head but a third may appear between them, generally shorter but occasionally longer than the others, and it can be observed that the latter are diminished in proportion as the middle one is increased. This affords a partial explanation of the cases in which one sex carries a pair of horns and the other a single one, as in *O. sagittarius,* Plate 5, figs. 13 & 14.

Apart from species closely related to one another, we may say of horned beetles in general that large horns are found in large insects, medium-sized horns in medium-sized insects, small horns in small insects and none in the smallest. But small species may have long horns. For example, the Indian *Onthophagus mopsus,* a beetle smaller than a dried pea, has a very long horn but it is of hair-like fineness.

If, bearing in mind the counterbalancing changes described above, we regard the total armament in each case we may enunciate, as a general rule for horned males of different but closely related species, that, *when*

104

arranged in the order of their average size, the total bulk of the horns (and equivalent outgrowths, if present) of specimens of full size will be found to increase in the same order but at a more rapid rate.

Thus amongst the different members of a species and amongst different species closely related we find the size of the horns controlled by the same mathematical principles, the effect of which must be automatically that a progressive reduction in the size of a horned insect, if continued long enough, will result in the disappearance of its horns, while persistent increase in size will produce in them a more and more disproportionate extravagance. The astonishing inconstancy of body-size in the individual specimens of many such species is repeated in the diversity of the species constituting the genera and such inconstancy is most apparent in those in which gigantic proportions have been attained.

The Goliath-beetles illustrate the error which may result from taking length as a measure of size. The males of all the species carry a horn upon the head. In *Goliathus Fornasinii,* which reaches scarcely half the size of the largest species of the genus, the horn stands up freely from the middle of the head and tapers to a slender terminal fork (see Plate 11, fig. 2), while in the larger forms the much shorter but more massive double appendage rising from the front of the head has a broad base carried from front to back, as in *G. albosignatus,* fig. 3. Although in the smaller form the length is greater, if a means could be devised of arriving at the actual bulk of the appendage in each species, large or small, we should no doubt find a regular progression from the latter to the former.

The variation exhibited by the horns in different individuals of a species is not always confined to the degree of their development. In discussing mandible-horns I have described cases in which two different and unconnected forms occur in different males of the same species. This also occurs in outgrowth-horns and one instance has already been mentioned, viz., the Tropical American *Enema Pan,* of which males with a long pointed head-horn and a double thoracic horn and others with the head-horn forked and the thoracic horn simple may be found together. As in Stag-beetles, the two phases may even, in rare instances, be found upon opposite sides of a single specimen. The male of another Tropical American insect, *Megaceras Jason,* of similar size to *Enema Pan,* has a horn

105

upon its head and another, probably the most massive known in any beetle, upon its back. The head-horn in a large specimen is forked at the end, while almost the whole upper surface of the thorax is raised into a huge hump, half as wide at the top as the insect itself, with two blunt pommel-like arms in front. In a small specimen the head has only a short undivided horn and the great hump is absent and represented only by two little projecting points near the middle of the back. From this stage we can trace in successively larger specimens the lengthening of the head-horn and development of the hump, until the latter is half as wide as in the largest examples. Here an unbridged gap appears and no intermediate specimens link the huge-humped giants with the more ordinary-looking males. In a large collection found in one place in Ecuador 16 males have a simple head-horn and at the most a moderate-sized hump and 16 have a forked head-horn and an enormous hump. One only, belonging otherwise to the smaller phase, has the horn almost forked and one has the right side of the hump large and the left side small thus combining the two forms, like the rare specimens of Stag-beetles I have described, in which two types of jaws are similarly combined in one individual.

M. Jason and another species, *M. Stuebeli,* are the largest of the numerous forms which compose the genus *Megaceras*. When *M. Stuebeli* is seen in sufficient numbers it is likely that it will also be found to have two male phases, for the tendency to split up in this way seems to be one to which giant beetles are especially liable. It is peculiar also to the largest males of a species. All small and medium-sized specimens, no matter how varied their horns, are found to fall into line with one phase and only large ones possess a second phase.

Different male phases are sometimes confined to different parts of the region inhabited by a species. For example, *Dynastes Gideon,* which ranges from India to Northern Australia, has always a tooth upon the head-horn in large males from India and Malaya but not in those inhabiting the Papuan region and Australia. In small specimens there is no tooth nor do females differ in any way. I have already mentioned a similar difference in large males of *Dynastes Hercules* from the American continent and the West Indian islands respectively. The most striking case of this kind is that of *Trichogomphus lunicollis,* another great Dynastid, with horns upon both the head and thorax, which inhabits the Malay Peninsula, Sumatra and Borneo. In the Peninsula and Sumatra large males bear upon the back a great hump with two arms which spread out laterally. In Borneo this, instead of spreading laterally, is drawn forward as a tapering horn which is forked at the end. The head-horn, in Bornean specimens, is at first straight and then bent back and bears a large process behind; in the other phase it curves gently and bears only a small tooth behind. These differences again are to be found only in full-sized males and all other specimens are of the same type wherever found.

The genus *Heliocopris,* containing the gigantic alllies in Africa and the East of the British *Copris,* exhibits a number of striking examples of horn-variability. One of the best of these is *Heliocopris gigas,* which inhabits many parts of Africa and continental Asia. The large males, which may be almost the size of a cricket-ball, bear two great horns rising from the sides of the head, a massive process projecting from the middle of the thorax and another from each of its front angles. In examples from Southern Africa the dorsal horn is rounded at the end, while the produced front angles appear as though cut off at the tip. Specimens from India have the dorsal horn squared, instead of rounded, at the end and the lateral processes are sharp-pointed. A gradual change can be traced from region to region, Arabian and Egyptian specimens being intermediate between the Indian and African races. But the differences are only found in well-developed males. In small ones head-horns and thoracic horns, median and lateral, all vanish, but upon the head, instead of horns, appears a transverse ridge, of which the large males have no trace but which corresponds to a ridge always found in the female. There is an inter-mediate male phase in which two teeth occupy the place of the ridge, while two more represent the lateral head-horns, making a row of four teeth across the head. All these various phases appear quite distinct until, by comparison of long series of specimens, it is seen that they are only separate links in a continuous chain. Exactly similar phases are found in the related Asiatic *Heliocopris dominus,* of which large males have two long horns upon the head, smaller ones four short horns and the smallest only a rectangular ridge. The actual form of these processes is evidently of little consequence.

To sum up — we find that, as noted in the case of mandible-horns, the most conspicuous fact emerging from the study of these outgrowths is their astonishing inconstancy. All the cases in which it is possible to explain them as apparently serving practical purposes are amongst those which have not attained more than a very moderate size. The vast majority, including all those which seem to us extravagant in their degree of development, defy every attempt to account for them as serving any purpose whatever. The greater the degree of development attained the less constancy is to be found and this can only point to the conclusion that no precise adaptation for the performance of any important function has occurred. Considerations of this kind led DARWIN to conclude that these remarkable structures were to be regarded chiefly as ornaments serving to attract the opposite sex and his theory will be discussed in another chapter.

CHAPTER 7.

COMPARISON WITH HIGHER ANIMALS

No-one can fail to notice the similarity between the horns of many beetles and those of higher animals. The stag's branching antlers, the long slender horns of the antelope, the short recurved ones of the chamois, the erect nasal appendage of the rhinoceros, all have their counterparts among the beetles. But the parallelism is much deeper than mere similarity. In the Coleoptera, as in mammals, the term "horn" is used as a convenient name for a variety of structures which are really very different but have certain characteristics in common and these are the same in both. The antler-like jaws of the stag-beetle, the nasal process of the Rhinoceros-beetle and the great dorsal projection of the Hercules-beetle are different things, as the periodically shed antlers of the stag, the permanent horns of cattle and, still more, the nasal horns of rhinoceroses and the tusks of elephants, are different, but certain features are common to them all. In horned insects, both those in which the adult condition is reached by gradual stages, as in bugs such as the fantastically horned Membracidae and the genus *Ceratocoris,* as well as in beetles in which, like butterflies and moths, it is attained abruptly, horns are at first absent and their appearance is a mark of approaching maturity, as it is in quadrupeds, although in these the approach is more gradual. Horns are also predominantly a mark of the male both in insect and mammal. Females may be quite hornless, as in nearly all deer, in the Goliath and Hercules-beetles, etc.; they may have horns but in a diminished form, as in many cattle, sheep and goats and some beetles; or rather rare exceptions, like the reindeer and such beetles as *Phanaeus lancifer* may be found, in which both sexes have a highly developed and practically equal armament. A greater development in the female than in the male is almost unknown and in the very few beetles in which it does occur a compensating outgrowth appears elsewhere in the male. In *Onitis tridens* and *Castelnaui*

109

although the females alone have horns the males have a great process beneath the body which is absent in the female.

But the most important feature shared in common is the relation between horn-bearing and size. In mammals and insects alike the finest examples of horns are found, not amongst the fiercest and most pugnacious of creatures, but amongst the bulkiest, for, as we include the jaws of the great stag-beetles as well as the long horns of the Hercules-beetles, so not only the spreading antlers of the great deer and the horns of the rhinoceros but also the tusks of the elephant must be taken into account. Not the carnivorous but the herbivorous mammals and not the fierce predacious beetles but the stolid, pacific and often social kinds are distinguished in this way. It is the largest representatives of any group that show the appendages at their fullest development while small, though nearly related forms are often quite destitute of them. It is the largest individuals of any species that carry the largest horns and, most important of all, these are not only absolutely larger but much larger in proportion to the size of the individual. In mammal and beetle alike there is a similar mathematical relation between the size of the appendages and that of the animal bearing them. In the mammal, although owing to the more steady and continuous growth they may become visible in an earlier stage, as maturity approaches they develop more rapidly than the bodily frame and their size continues to change according to the size and weight of the individual. This is most clearly seen in the case of the deer, whose antlers are periodically shed and renewed, their size increasing regularly until the bearer attains his maximum size and weight and declining as these decline. In the insect, in which maturity is attained suddenly on the last casting of the skin, after which there is no further growth, the size of the armament also depends upon that of the individual, which often differs extremely in different examples of the same kind.

Again, in insects and mammals alike, while the structures may often serve a useful purpose, it seems that in many cases their development has proceeded to fantastic lengths which bear no relation to practical purposes and must inevitably be a cause of embarrassment, if no worse. It is said that the ponderous horns of large male individuals of Marco Polo's Sheep (*Ovis Ammon*) so seriously reduce their agility that they fall easy

110

victims to wolves. The antlers of the stag are an undoubted impediment in wooded country and, in the contests which are frequent between the males, may become interlocked so that they cannot be separated, so causing the death by exhaustion of both animals.

In Africa it has been noticed that contending males of the Elephant dung-beetle (*Heliocopris gigas*) are sometimes found dead with horns interlocked and the same has been reported of the South American beetle *Diloboderus Abderus*.

Geology shows abundant evidence of the steadily increasing size from age to age of animals of many kinds and all the giant forms, vertebrate and invertebrate, are the descendants of smaller ancestors. Deer and other horned animals can be traced back in the geological record to small and hornless ancestors. If this process of gradually increasing size involves, as it appears to do, a constantly greater increase in the proportionate size of the burden to be borne by horned species, it must at last reach a stage at which that burden is no longer compatible with success in the struggle for existence. Evolutionary changes are unfortunately too slow to be perceived by any observations we can make and there is no means of judging whether this process is still in operation in any particular case, but it is at least highly probable that there are cases in which it is still operating. The great so-called Irish Elk (*Alce gigantea*), which vanished from the earth shortly before the dawn of human history, once ranged widely through northern Europe and Asia. Its immense antlers had a spread of as much as eleven feet and it is considered not improbable that its disappearance may have been hastened by these having become too great an encumbrance. The common Red Deer (*Cervus elaphus*), once contemporary with the Irish Elk, was at that earlier period, as its remains show, a larger animal than now with much heavier antlers, but the larger race failed to endure conditions which its smaller descendants have survived and to this reversal of the evolutionary process the survival of the species may be attributed.

The Mammoth and other extinct kinds of elephants, which existed in great numbers about the same time as *Alce gigantea* and ranged over a far larger part of the world than their representatives at the present time, bore tusks more burdensome than those of the latter and this may perhaps have contributed to their extinction. A single tusk of the African elephant may measure 10 or 11 feet in length and a pair may weigh as much as 450 lbs, but those of the great American *Elephas imperator* are said to have reached a length of 16 feet. The late R. BLAYNEY PERCIVAL, whose post as Game Warden afforded him many opportunities of observing African elephants under natural conditions, gave a moving account of the fatigue imposed upon the large bulls by their heavy burden and the necessity of relieving the overstrained muscles by periodically resting their tusks. The disappearance of species still more heavily weighted, whether or not there were other contributory causes, could be accounted for by supposing that the continuing increase of size at last resulted in handicapping them to an unendurable extent. The fact that the tusks of the Indian elephant are sometimes absent in the male and generally in the female is believed to indicate that they are in process of disappearance, but there is no reason to suppose that the animal is at any disadvantage on that account.

In earlier geological times gigantic animals belonging to various groups and bearing horns of large size have disappeared and left no descendants. The antler-carrying short-legged giraffes (*Sivatherium*) of Europe and Asia in the Pleistocene age and at an earlier time the huge six-horned American *Dinoceras* and the Egyptian beast, *Arsinoitherium,* with immense double horns of a kind entirely different from any now existing, were all the last of their respective lines of descent, as were certain far earlier Dinosaurian reptiles (*Triceratops,* etc.) also furnished with horns. Smaller forms of life have over and over again been able to survive difficulties that have overwhelmed the giants. As these become extinct, small forms of some different group may be increasing very gradually in size at the same time, perhaps in their turn to become giants and pass away from similar causes.

Amongst beetles, as with vertebrate animals, the giants in any group have a special tendency to acquire unwieldy appendages and there is no doubt that they also are the present-day descendants of smaller ancestors and that here also a gradual increase of size, continued through long ages, has been accompanied by a more rapid increase in their armature. No prolonged examination of such astonishing forms as the Hercules-

112

beetles is needed to produce conviction that any further increase of size, with its consequence, a still more disproportionate growth of the horns, must of necessity result in destroying the balance of the insect, make flight impossible and ultimately bring about extinction. We may reasonably assume that in past times other gigantic forms such as these have existed and disappeared. Unfortunately we know little of the geological history of the Lamellicorn beetles but it is probable that horns were borne by many beetles of past ages long before the appearance of the warm-blooded animals in which they are most familiar to us, just as flight was practised by insects ages before birds had made their appearance in the world.

But these appendages are not peculiar to mammals and insects. Not only were they borne by some extinct reptiles, they are found in some still living, chameleons and snakes, as well as in birds and fishes, nearly always showing the characteristics of other so-called secondary sexual features, a special association with the male sex and a marked tendency to assume curious and fantastic forms.

Since in familiar cases it is known that the horns serve as weapons it is a natural assumption that such is their function in all cases, but this is certainly not so, for comparatively few show any actual adaptation for that purpose. Amongst the almost infinite variety of beetle horns very few show any suitability for such use. Even amongst mammals the defects of some of the forms met with, regarded as weapons, are obvious. It is usual to find in animals of whatever kind that organs serving a particular purpose are adapted with the greatest exactitude for the most efficient performance of that function and that variations of form are related to variations in the manner of life; but it can hardly be supposed that, as for example in the antelopes, the horns of one point upward, those of another of almost the same form and habits point downward, of a third backward and a fourth forward, in order to gain the highest degree of effectiveness as weapons. Does not the variety of forms indicate that such effectiveness cannot in all cases be of great importance? It is said that with its straight slender horns, which may be more than a yard long, the Gemsbuck has been known to kill a lion. It is notoriously the most formidable of all the antelopes, but its horns are the simplest in form. Although there are many larger and more powerful species of the tribe,

113

provided with horns of more elaborate design, none can use them with such good effect and their elaboration seems, instead of improving them in this respect, to have resulted in diminishing their efficiency as weapons. The massive horns of many of the goat tribe, unwieldy as they are, are evidently less fitted for combat than smaller weapons would be and it has already been mentioned that those of the wild sheep (*Ovis Ammon*) may be a dangerous handicap to it.

The ultimate degree of extravagance in mammalian horns is reached by the antlers of some of the great deer. It must be apparent to all on reflection that the immense size and fantastic branching of those of the Reindeer, the Elk, in which they may measure six feet from tip to tip, and, still more, those of the extinct Irish deer, sometimes eleven feet across, cannot be regarded as increasing their practical usefulness and that, for animals inhabiting wooded country, as do many of these deer, the wide-spreading appendages can only be a handicap. It is of interest to find that the only kind of deer of which the female, as well as the male, is so encumbered is the one, the reindeer, whose normal habitat is the far north beyond the forest-line and that where the animal has extended its range into a wooded region of North America, a race has been produced in which the antlers are relatively shorter than those of the parent form. The inadequacy of antlers as means of defence is increased by the fact that they are shed and redeveloped periodically, so that for a considerable part of every year they are either absent or in a soft and delicate condition during the process of renewal. The well known fact that, owing to their elasticity, fighting stags sometimes get them so entangled as to be unable to separate them, resulting, in the wild state, in the miserable death of both, shows that as offensive weapons also they have undoubted short-comings.

In contests between rival males of a species, horns, when present, from their situation, may be involved, almost of necessity, but such contests probably are nearly universal and the result, — victory for the boldest and most vigorous — is the same for horned and hornless alike. The use of horns in such contests therefore seems to have no special advantage for the species. Fighting animals of all sorts use whatever part of the body will best serve the purpose. Whether they employ horns, hoofs, teeth or fists, these must have acquired the necessary degree of effectiveness before they could be brought into service. The stages by which hoofs, teeth and fists have reached their present condition can be traced far back with considerable certainty, but the origin of horns remains much more obscure. We have found that in groups of animals so remotely related as beetles and quadrupeds they show the same characteristics and the question of their origin and significance can therefore be regarded as a single problem. Since horns are only one type of the great range of features known as secondary sexual characters, any light that can be thrown upon the problem will be a contribution to the understanding of the whole multitude of puzzling and often fantastic features specially associated with the male sex throughout the animal kingdom, such as the song of the nightingale, the vivid colours and exaggerated plumes of pheasants and birds of Paradise, the combs and wattles of other birds, the crests of lizards and newts, and innumerable others.

CHAPTER 8.

DARWIN'S THEORY OF SEXUAL SELECTION

If their use as weapons is inadequate to account for the many strange horn-forms met with amongst mammals, the inadequacy of such an explanation is much more apparent when we proceed to consider the horned creatures of many other kinds and especially the insects, which include more numerous examples than all the other groups together. Of these the great majority are beetles. A notable attribute of the horns of beetles, as of other animals, is their liability to attain fantastically disproportionate size. In this they resemble other features distinctive of male animals for which no practical purpose can be discerned, such as the tusks of the elephant and narwhal, the remarkably elongated plumes of various male birds and fin-rays of certain fishes, the immense claws of some crabs, etc. Such extravagances are not found in any ordinary organs of which the function is apparent. The real significance of all these very remarkable and extraordinary male features is a biological problem which has attracted the attention of many enquirers — and amongst them one of the greatest of all enquirers, CHARLES DARWIN, who devoted the greater part of his book, "The Descent of Man", to the consideration of it. Impressed by the impossibility of regarding many structures of this kind as practically useful, and so of accounting for their development by the principle of the survival of the fittest, DARWIN reached the conclusion that they could best be explained as ornaments serving to attract the opposite sex. He considered that the surprising development of many such male features could be produced by the females exercising a preference for those males exhibiting them in the highest degree and rejecting those less splendidly endowed, so that such ornaments were transmitted to succeeding generations in ever-increasing splendour. Applying this to beetle-horns, he remarks "It is probable that the great horns possessed by the males of many Lamellicorns and some other beetles have been acquired as ornaments." (Descent of Man, 1910 ed., p. 502)

116

and, in further explanation of the theory, as applied to insects — "Sexual selection implies that the more attractive individuals are preferred by the opposite sex; and as with insects, when the sexes differ, it is the male which, with some rare exceptions, is the more ornamented, and departs more from the type to which the species belongs; and as it is the male which searches eagerly for the female, we must suppose that the females habitually or occasionally prefer the more beautiful males, and that these have thus acquired their beauty... Judging from what we know of the perceptive powers and affections of various insects, there is no antecedent improbability in sexual selection having come largely into play; but we have as yet no direct evidence on this head, and some facts are opposed to the belief. Nevertheless when we see many males pursuing the same female, we can hardly believe that the pairing is left to blind chance and is not influenced by the gorgeous colours or other ornaments with which the male is decorated" (*op. cit.,* p. 504). This passage affords an admirable example of DARWIN'S always fair and moderate statement of his case. Covering the entire range of the animal and vegetable kingdoms, as the immense field of his enquiries necessitated, he was dependent for the majority of his facts upon the knowledge of others and 70 years ago what was known of the perceptive powers and affections of insects was very little, for before DARWIN himself showed the way, the number of naturalists engaged in exact investigations was small and, in enquiries relating to insects, very small.

In other statements DARWIN gave a rather wider application to his theory than is contained in the definition quoted above and included, besides the preference of the females for the most attractive males, selection by contests between males of those armed with the best weapons. This however is an entirely different matter, the evidence for which is much easier both to obtain and to interpret. That contests between rival males occur in many kinds of animals is certain but the outcome, depending upon the strength and skill of the contestants, as well as upon the efficiency of their weapons, is clearly to be regarded as the survival of the fittest and one of the modes of operation of Natural Selection. The problem with which DARWIN found himself confronted was to account for all those varied features found in male animals which very evidently do not make

117

for greater success in combat. Sexual Selection, according to his own conception, was a process supplementary to Natural Selection and not a part of it. It is therefore better to restrict the term, as he has done in the passage quoted, to preference exercised by one sex for the best adorned or in other respects most attractive individuals of the opposite sex. The implications of this theory are various. It requires the possession of highly developed vision with power to appreciate slight differences of size, shape and colour, together with a certain degree of aesthetic susceptibility and also the persistence through innumerable generations of the same unvarying standard of excellence. The first requisite no doubt exists at least in the most advanced vertebrate animals; whether they possess the others is less certain. We can see evidences of some aesthetic appreciation in a few animals, for example in monkeys and bower-birds, but it is of a very elementary kind. As THOMSON and GEDDES remark (The Evolution of Sex, 1889, p. 28) "Even among birds, if we take those unmistakable hints of real awakening of the aesthetic sense which are exhibited by the Australian bower-bird or by the common jackdaw in its fondness for bright objects, how very rude is this taste compared with the critical examinations of infinitesimal variations of plumage on which Darwin relies." As to insects, the existence of such a highly developed faculty is certainly not easier to accept, but this may perhaps be regarded as immaterial, for the keen vision demanded by the theory is in critical cases absent.

In 1881 a German author, W. VON REICHENAU, in an essay "On the Origin of secondary male sex-characters, especially in the Lamellicorn beetles" (Kosmos, 10, p. 172) brought forward many arguments for rejecting sexual selection as applied to insects and maintained as the rule throughout the group that "the female belongs to the male who first finds her" and exercises no choice whatever; in other words that marriage by capture is the universal rule. Reichenau dealt especially with the horned beetles and, amongst other arguments, called attention to the essential difference between the very delicate protected and mobile eyes by which we and the higher animals are able to perceive slight shades of difference and the hard immovable eyes of insects. Although he did not elaborate the point, it is in reality a fatal weakness in DARWIN'S theory so far as it relates to insects, whose vision is of an entirely different kind from that of vertebrate animals. Its comparative unimportance to them is shown by the extent to which the most complicated operations are often performed in complete darkness. All the work of the hive in bees, the orderly coming and going of the workers, the building of the comb, the tending of the brood, all the corresponding operations of ants in the nest, the labours of the Ambrosia beetles and the lengthy processes of nidification in *Copris, Geotrupes* and countless other insects, are carried on in total darkness by means of their delicate senses of touch and smell, without the assistance of sight. Some, otherwise highly organized and very agile insects, are quite blind and there is even a blind stag-beetle, *Vinsonella caeca,* in the island of Mauritius, the male of which, as usual in the family, has rather larger jaws than those of the female, but, since she is unable to see them, this must be attributed to another cause than sexual selection.

The degree of vision enjoyed by insects differs greatly in different cases but it is certainly always very unlike that which we ourselves possess. As it is easy to prove experimentally, the eyes of an insect are very sensitive to changes in the light that falls upon them but, provided it is not interposed between them and the source of light, an object quite a short distance away is often not perceived. Many people must have noticed the mistakes made by butterflies and other insects attracted by artificial flowers or coloured representations and the very close and repeated examinations necessary before they are convinced of their mistake ; yet the vision of butterflies is comparatively good. Certain extraordinary kinds of "praying Mantis", (*Hymenopus,* etc.), which, when lying in wait for their prey, bear a resemblance to flowers, take advantage of this weakness and devour the unfortunate creatures that alight upon them by mistake. Careful experiments have been made to test the quality of butterfly vision. DR ELTRINGHAM found that the British Pearl-bordered Fritillary (*Argynnis euphrosyne*) appeared able to see another specimen of its kind at about a distance of a yard, but could distinguish head from tail only at from 2 to 4 inches. The ordinary eyes of insects, it is generally known, are compound. Some have, in addition, one or more simple eyes (ocelli), but these simple eyes probably only perceive light and form no definite image. They are not found in the horned beetles with which we are concerned.

119

The compound eyes generally consist of a very large number of separate elements (transparent rods and lenses), each with its own facet, which receives the light falling upon it from a small part of the field of vision, and each connected by nerve-fibres with the brain. We must conclude that the result is a mosaic pattern, finer or coarser according to the number of separate elements composing the organs. Generally there is a more or less hemispherical mass of facets on each side of the head, often numbering many thousands. The hemispherical shape provides the largest field of vision, collecting the light rays from as many directions as possible, and the two hemispheres may occupy the greater part of the whole head, as in many bees and two-winged flies, butterflies and moths. The compound eye of the house-fly contains about 4000 separate facets, those of butterflies and moths from 12,000 to 20,000. The visual powers of different insects therefore differ enormously. Generally speaking the most active insects have the best sight and some dragonflies, which are very swift and hunt their prey upon the wing, have as many as 28,000 facets in each eye. Most beetles are much less well equipped. Some, such as the Tiger-beetles, which run and fly swiftly, have large eyes, prominently situated at the sides of the head to give a wide and clear field of vision, which is assisted by the free attachment of the head. Each eye may have about 4000 facets and there is no doubt that their sight, although inferior to that of bees and butterflies, is good compared with that of most other beetles. If we examine the eyes in any of the main groups containing horned beetles we shall be obliged to conclude that their vision is, on the whole, much inferior to that of the Tiger-beetle. A few, such as the Rose-beetles, have, it is true, hemispherical and prominently placed eyes, although smaller and less well placed than the Tiger-beetles', which may contain as many as 4000 facets. These are very active day-flying insects, like the Tiger-beetles, and like them also amongst the best-sighted. But horned Rose-beetles are quite a small company and the vast majority of horned, as of all other beetles, are less lively creatures, which generally remain in concealment during the day and prefer darkness or twilight for their activities. In consequence sight is of less importance to them than their other senses. Its relative unimportance is shown by the way in which the

120

eyes are often sunk into the head, as in the mole, which also has little use for them.

In the common Stag-beetle, *Lucanus cervus,* the eye is very small, the facets numbering about 2000, most of them beneath the head, which is dilated in front. Objects upon the ground or partly behind the head are therefore best in view. The antenna, the most important sensory organ, is attached to the head immediately in front of the eye and is in constant motion there, a further obstruction to forward vision. In many other Stag-beetles, such as those of the genus *Calcodes,* each eye is completely divided into two halves, the larger half situated beneath the head for surveying the ground and the smaller half sunk in the upper surface facing the sky, but neither part affording any general outlook upon the world. Indeed it is not possible, so far as we can judge from their eyes, for any horned beetle to form a clear picture, as DARWIN supposed, of another. In the Hercules and Rhinoceros-beetles (Dynastinae), the Minotaur and other Geotrupinae, the eyes are similarly divided by a ridge into upper and lower halves and often consist of less than 2000 facets. The Scarabaeinae, which include the most numerous horned forms of all, have a broad shovel-like head, beneath which the eyes are placed, with only a tiny part of each directed outwards. In the curious *Onthophagus imperator,* Plate 5, figs. 11 & 12, of which male and female have each great horns of a different pattern, the upper part of the eye contains about 250 facets and the lower about 1000, while the related *O. pyramidalis* has not more than about 800 in all. I have already mentioned the blind Stag-beetle of Mauritius, *Vinsonella caeca,* in which every trace of eyes has disappeared. In India is found a much larger Stag-beetle, *Aulacostethus Archeri,* little smaller than the great *Lucanus cervus* and possessing large horns in the male but with eyes reduced to such tiny vestiges that it must be almost completely blind.

In that group of Longicorn beetles which, after the Stag-beetles, is most notable for mandible-horned males and includes the immense *Macrodontia* (Plate 3), the sight is little, if at all, better than that of the Lamellicorn beetles. The eyes are often exceptionally coarsely faceted, indicating very blurred vision. This coarseness of the outer eye usually indicates that the insects are only active during the hours of darkness,

121

when the other senses must be chiefly relied upon. Longicorn eyes are sometimes fairly large but, instead of forming hemispheres, are generally more or less kidney-shaped, being cut away in front for the attachment of the antennae, which, in constant movement directly before any lenses with a forward outlook, must still further blur the picture. The total number of lenses in the very narrow eye of the big-jawed Psalidognathi is only 700 or 800 and in another long-jawed genus, *Mallodon,* as low as 600.

These examples will serve to illustrate the fact that the groups most conspicuous for the number of horned beetles they include are all far inferior in visual power to those insects we can consider as having fairly good sight. The great assemblage which contains most of all, the Scarabaeinae, is perhaps the one in which the faculty is at its worst, while the next most important, the Lucanidae, is little better off. It need not be supposed that there is any close relation between the possession of horns and poor sight, since in all the groups hornless as well as horned forms are found. The real relation is no doubt with the manner of life, the best sight being possessed by the most agile insects, those especially which hunt for living prey. None of the beetles here dealt with are predacious and most of them are comparatively slow in their movements. Many of them pass much of their lives in burrows below the ground, in logs or tree-trunks. This is especially the case with the Scarabaeinae, many of which are occupied for lengthy periods with underground tasks, working in total darkness.

The truth is that for most, if not all, insects the most important and best developed sense is not sight but smell, which reaches a degree of perfection with which the faculty in ourselves to which we apply the term is scarcely at all comparable. This sense in insects is located in the feelers or antennae, which, except when the insect is at rest, are always extended and in motion, "feeling" the air for the sensations which chiefly control its actions. It is by this sense that insects are able to locate and recognize their food-substances, as well as others of their species. The astonishing efficiency of their olfactory sense has been already illustrated by the power of *Bolboceras* to find the fungus, hidden beneath the soil, upon which it feeds, and still better by the tracking down of the inert female of *Pachypus cornutus* by the male, although similarly hidden under the ground.

Experiments with ants have shown in a striking way how small a part sight plays in mutual recognition amongst those insects. Worker ants introduced by any means into the nest of another species are invariably treated with deadly animosity but it has been found by MR B. D. W. MORLEY (*Proc. R. Ent. Soc. Lond.*, 15, 1940, p. 103) that workers of different species, irrespective of colour, size and shape, will at once settle down harmoniously together if first treated in such a way as to impart to all the odour characteristic of one of the species. The fan-like structure of the antenna in Lamellicorn beetles, by increasing the area of sensitive surface, has increased the delicacy of the olfactory sense and this is often especially developed in the males, whose antennae are very frequently superior to those of the females; but for those of the latter to show a greater development than those of the male is quite unknown, showing that in the meeting of the sexes the male and not the female is the active partner.

Since DARWIN admitted that some facts were opposed to his theory, noting, for example, the habitual indifference shown to the males by female silk-moths, it cannot be doubted that, had he become aware of the facts stated here concerning the vision of horned beetles, he would have recognized them as destructive of his theory of Sexual Selection, so far at least as it applies to those insects, for if, as he believed, certain female insects were able to compare their various suitors and select from among them those with the best physical attractions, such insects must surely be found among those with the best and not the worst quality of vision. Not only does all the evidence indicate indifference on the part of female insects but we find, in the case of those under consideration, that the differences so manifest to ourselves in the males cannot be perceived at all by the females.

ALFRED RUSSEL WALLACE, the great naturalist who shares with DARWIN the fame of having discovered the vital principle of Natural Selection, in his book "Tropical Nature", published in 1878, and afterwards in his splendid exposition of the doctrine which, with fine generosity he called "Darwinism", rejected Sexual Selection for two reasons, first that, as DARWIN himself admitted, observed facts gave it no support and secondly that, even if female birds, butterflies, beetles and other creatures under natural conditions were found to select for their mates the most decorative males, the result would not be as DARWIN supposed. WALLACE based his argument primarily upon the case of birds, in which the well known and remarkable facts of the display by the males of their special attractions certainly make DARWIN'S supposition most probable. He says "Natural Selection, as we have seen in our earlier chapters, acts perpetually and on an enormous scale in weeding out the "unfit" at every stage of existence and preserving only those which are in all respects the very best. Each year only a small percentage of young birds survive to take the place of the old birds which die; and the survivors will be those which are best able to maintain existence from the egg onwards, an important factor being that their parents should be well able to feed and protect them, while they themselves must in turn be equally able to feed and protect their own offspring. Now this extremely rigid action of Natural Selection must render any attempt to select mere ornament utterly nugatory unless the most ornamented always coincide with the "fittest" in every other respect; while, if they do so coincide, then any selection of ornament is altogether superfluous. If the most brilliantly coloured and fullest plumaged males are n o t the most healthy and vigorous, have n o t the best instincts for the proper construction and concealment of the nest and for the care and protection of the young they are certainly not the fittest and will not survive or be the parents of survivors. If, on the other hand, there is generally this correlation — if, as has been here argued, ornament is the natural product and direct outcome of super- abundant health and vigour, then no other mode of selection is needed to account for the presence of such ornament. The action of Natural Selection does not indeed disprove the existence of female selection of ornament as ornament but it renders it entirely ineffective; and, as the direct evidence for any such female selection is almost nil, while the objections to it are certainly weighty, there can be no longer any reason for upholding a theory which was provisionally useful in calling attention to a most curious and suggestive body of facts but which is no longer tenable". (Darwinism, 1889, p. 295). This passage has been criticized

124

by CUNNINGHAM (Sexual Dimorphism in the Animal Kingdom, 1900, p. 25) "(WALLACE'S) objection is not a very powerful one for it was obvious from the first that the female could only exercise her choice among the individuals which had survived in the struggle for existence", but this misses the point of WALLACE'S argument, the extreme improbability of the perpetuation of mere ornament unless invariably accompanied, as with good reason he believed it to be, with practically useful qualities. When we consider the numerous instances of birds of which the female enjoys the protection of concealing colour, while the male is conspicuously, and therefore with regard to his foes disadvantageously, coloured, we can hardly fail to agree that, unless invariably accompanied by some practical advantage, such as "superabundant health and vigour" to outweigh the disadvantage, the ornament would not have been perpetuated. During the fifty years which have passed since "Darwinism" was published our knowledge both of the habits of birds and the significance of colour and ornament has considerably increased but as to the existence of "female selection of ornament as ornament" WALLACE'S statement that "direct evidence is almost nil" remains unshaken. Little support can be found to-day for Sexual Selection in DARWIN'S sense and DR JULIAN HUXLEY reviewing "DARWIN'S theory of Sexual Selection in the light of recent research" (American Naturalist, 72, 1938) has suggested that the term should be allowed to drop out of use. Among the few expert authorities who have given the theory their support are MR & MRS PECKHAM, who, after making many exact observations of the courtship of spiders, and the display, curiously resembling that of birds, made by many of the males in the presence of the female, pronounced it likely that the latter may exercise a preference for one male rather than another. "When MR WALLACE denied that the choice of the female was influenced by ornamentation (not because this would not explain the facts but because he disbelieved in any such predilection on her part) he was left without any explanation of the very important fact that colour and ornament in birds, insects and spiders almost always appear in certain parts of the body... What is the difficulty in supposing that females have selected one ornamental variation rather than another and that, as the generations passed, this process being cumulative, all the sexual colour and ornament

125

that we find in Nature have resulted?" (G. & E. PECKHAM, Sexual Selection in Spiders of the family Attidae, 1890, p. 134). But despite their very precise observation of the spiders' behaviour, MR & MRS PECKHAM were themselves unable to find any evidence of selection by the females. "It is true that we have no direct proof that the females select the more beautiful males" (*op. cit.,* p. 146). In more recent times MESSRS BRISTOWE & LOCKET, reviewing the same remarkable phenomena in the light of their own observations, have put forward another explanation of them. They believe that the sight of spiders is poor and that they are unable to distinguish clearly objects by its means even a short distance away. The male locates the female by scent, the olfactory organs being at the tips of the legs, but, owing to the shortsightedness of the female, his approach is attended by the risk that he may be mistaken for legitimate prey, pounced upon and killed immediately on coming near. His peculiar antics and display of his distinctive marks may therefore serve the important purpose of distinguishing him from other creatures more legitimately serving as prey. (*Proc. Zool. Soc.,* 1926, p. 317).

The close relation between the size of beetles and other animals and their horns has been repeatedly mentioned here and increasing size, by putting an animal just out of reach of some of its enemies, gives it an obvious advantage.

Various other suggestions have been put forward to account for beetle-horns. WALLACE, in "Natural Selection and Tropical Nature", 1895, p. 372, considered it possible that they might be protective. Since the males are more in the habit of flying about than the females particularly at dusk, they would be liable to attack by such birds as owls and goatsuckers and the horny outgrowths, like the sharp spines of some wingless ants and other insects, would make it difficult for the birds to swallow them. It has been shown here that the outgrowths have been adapted to serve a variety of purposes and this may be one, as, for example, in such species as the Australian *Onthophagus ferox,* in which horns have been acquired by both sexes and take the form of strong pointed spines directed straight outwards. This explanation is not capable of wide application, however, or we should find such a type occurring more frequently. When male and female beetles have similar horns they are usually of a quite different form.

Another explanation has been put forward in "The meaning of Animal Colour & Adornment", 1933, by R. W. G. HINGSTON. This is that the horns of stags and beetles, the tusks of elephants and such features in all their variety may be regarded as serving to produce an intimidating aspect and are thus means of protection from attack. "Some male stag-beetles have enormous jaws, extravagant far beyond physical needs... thus they possess the same attributes that characterise the antlers of stags... they are now mainly intimidating instruments." (p. 267). It has been mentioned as probable that some great beetles, like the extraordinary *Chiasognathus* described by DARWIN, obtain protection by taking advantage of the alarming aspect of their horns to intimidate enemies but the fact that the protection is not afforded to the female, which, especially during the critical operations accompanying egg-laying, has greater need of it than the more active male, renders its value to the species rather small. Moreover the earlier stages preceding the acquisition of an alarming aspect remain still unexplained and the frequently rapid increase of insects accidentally introduced into a new country shows clearly enough that the effective enemies of any insect are not the casual ones which might be deterred from attack by such means but those which have acquired a settled habit of preying upon it.

Yet another theory was advanced by J. T. CUNNINGHAM in "Sexual Dimorphism in the Animal Kingdom", viz., that all outgrowths of this kind are due to local mechanical stimulations. "Conspicuous as the excrescences of male Coleoptera are in themselves, they are even more extraordinary in their variability within the same species. Secondary sexual characters are often variable but there are no other animals in which such differences in the degree of their development in the same species are constantly found as in these two families of Coleoptera (i.e. Dynastidae & Lucanidae). In the Dynastides every degree of development may occur, from males which are almost similar to the females in everything, except that they are slightly larger, to males with an extreme development of horns and a size of body much greater than that of the female. In the Lucanidae very often the degrees of development fall into

127

two or more distinct stages, intermediate conditions being rare or want-ing.... Thus the males themselves are polymorphic. No theory based on selection has succeeded in explaining this remarkable state of things. DARWIN considered that the great range of variation supported the conclusion that the horns of Dynastides had been acquired as ornaments because, I suppose, in that case the imperfect development would be of little importance. But whether the structures are ornaments or weapons, if only the best developed are successful their perfection ought to be transmitted to their offspring, so that on this reasoning either sexual selection does not occur or it fails of its object." (*op. cit.,* p. 256). "Whether the excrescences are useful or not, it seems to me that their resemblance to similar unisexual developments in other families indicates that they have been produced by mechanical irritation, by blows of some kind." (p. 254). The author, that is, saw in horns, whether in beetles or other animals, weapons owing their origin and development to the habit of fighting and he found in this conclusion proof of the inheritance of acquired characters, "Unisexual characters... may, in many cases, be correctly described as excrescences and these excrescences are as truly the result, in the first instance, of mechanical or other irritation as a corn in the human epidermis" (p. 39)... "the direct effects of regularly recurrent stimulations are sooner or later developed by heredity" (p. 41).

Had the author possessed greater knowledge of the facts relating to the unisexual outgrowths of beetles, their occurrence in almost every possible situation, their extraordinary variability, not in the degree of development alone, and especially the occurrence of compensatory outgrowths, entirely incompatible with the idea of local stimulations, the existence of one-, two- and four-horned males of a single species (*Heliocopris dominus*), the replacement of a single horn in the male by two in the female (*Ontho-phagus sagittarius*) or of a horn on the head of the female by one on the sternum of the male (*Onitis tridens*, etc.), he would have realized how little support is afforded to his theory by the Coleoptera. Moreover he strangely failed to recognize the different effects of mechanical stimulation upon horny tissue like the exterior of a beetle and soft tissue such as the human epidermis. The effect upon the chitinous exterior substance of a beetle, as upon the mammalian horn, is easily seen to be, not further growth but wearing away, and it is also demonstrable that neither corns produced upon the human epidermis nor the effects of wear and tear upon the horny excrescences of stag, goat or beetle are ever transmitted to the succeeding generation. That interesting creature the Sacred *Scarabaeus,* which, since the time of the ancient Egyptians, who have left us so many representations of it, has occupied itself in rolling balls of dung along the ground, out of which the female fashions the cells in which its eggs are to be placed and its offspring grow to maturity, has a row of sharp horny processes at the front margin of its head and another row along the edge of each of its front legs. The ball is propelled by pushing it backward, the sharp processes of the head and forelegs being pressed against the ground. The effect of these regularly recurrent stimulations upon the processes is easily seen. They are soon blunted by the friction but, although this has been repeated for countless thousands of generations, the effect has not been developed by heredity and each young *Scarabaeus* on emerging from his empty cell has the projecting points of head and legs as sharp as those of its parents before their ball-rolling efforts began. The strange manner of life followed by these ball-making beetles has led to a very remarkable peculiarity. They have lost their front feet. The flexible tarsus, composed of several separate joints and terminated by a pair of claws, which forms the foot in adult insects, remains upon the four hinder legs of *Scarabaeus* but has disappeared completely from its forelegs. The jointed feet, with their claws, serve to give a secure foothold, which is of especial importance in climbing or clinging insects. But *Scarabaeus* does not climb. With its four hinder legs it clasps its precious ball and the claws and tarsi of these are indispensable for that purpose. For propelling the ball backward rigid, not flexible, implements are needed, as in propelling a punt, and the insect uses the toothed head and front tibiae in alternate strokes, changing its grasp of the ball by the four hinder tarsi as it rolls. In constructing the ball also, accomplished by pressure between the legs, the fore-tibiae are as essential as the butter-pats to a dairymaid, while the tarsi are not required at all. When the ball has been rolled sufficiently to serve the purpose, whatever that may be, — whether to compact it better, to expose it to sun and air or simply to convey it to a suitable place for concealment — a hole is scraped into which it is drawn for further manipulation and for this also the fore-tibiae are the essential implements. In all these proceedings not only are fore-feet unnecessary but it is apparent that they would constitute an actual encumbrance to the most expeditious performance of the complicated and difficult movements required. During the long ages in which these operations have been gradually perfected the superfluous feet have therefore vanished and the stages by which their disappearance has been brought about can be traced. It is not the result of inheritance from an ancestor whose fore-feet were accidentally amputated, for the result of amputation is not transmitted to a succeeding generation. The disappearance of the front tarsi in insects of this group seems to be a process still continuing; in many, although present and complete they have become reduced to very small dimensions and in some the shrinkage has progressed so far that they can only be detected with difficulty. It is not difficult to realize therefore that the process has been that called by DARWIN the Survival of the Fittest. Since any degree of diminution of these superfluous members lessened the hindrance caused by their presence in the course of the insects' operations, those individuals in each generation in which they were most reduced had an advantage over the rest, resulting in a steady reduction, continued in some cases and perhaps ultimately to be continued in others, to the point of total disappearance.

It seems then that the facts afford no support to the theory of Sexual Selection or that of stimulation by combat but that the explanation of horns is likely to be a more complex matter than has been supposed.

CHAPTER 9.

THE ORIGIN AND SIGNIFICANCE
OF BEETLE HORNS

It has been shown that a few types of beetle-horns can be explained as adaptations for certain purposes such as assistance in ascending a vertical shaft, in raising and expelling excavated material from the burrow or uniting more closely the head and thorax and so increasing the insect's lifting power. The adaptation in all such cases consists in the obliteration to a greater or lesser degree of the original horn-form. In certain other cases the armature may perhaps render its bearer a less easy prey to birds or other insectivorous creatures or reduce the risk of attack by simulating a menace that has no actual existence; but the fact that such protection is confined as a general rule to the male, although it is the female which, on account of her inferior agility and her greater exposure to attack when occupied in the critical operation of egg-laying, is in most need of protection, much reduces the value of this. It must be remembered also that it is the females of beetles, as of other insects, which ordinarily perform all the manifold operations of nidification and that they perform them without other implements than their jaws and feet. Indications of the adaptation of horns for such purposes are not numerous but it is certainly not easier to find evidences of adaptation for use as weapons. The extreme unsuitability of many types is very obvious, those borne by species of *Golofa* shown upon Plate 10, for example, or the slender trailing antlers of *Onthophagus gibbiramus* and *rangifer,* Pl. 5, figs. 7 & 9.

That contests between male beetles occur we know but that they are of a different or more serious character in horned than in hornless kinds there is no evidence at all. All the insects in question are well protected with body-armour and that the horns of any one of them have ever left a mark upon the body of another, with the possible exception of a few mandible-horned Lucanidae, there is no scrap of evidence. When a horn is borne in front of the body it can hardly fail to be involved in any

131

scuffle that occurs and MR W. BEEBE, in a very circumstantial account of fights between rival males of *Megasoma Actaeon* says that this beetle uses its cephalic horn in an effort to turn its opponent on to its back, but *M. Actaeon* is provided also with a pair of strong lateral horns and these appear to be of no use in the contest. The cephalic horn cannot but become involved but it is not at all apparent that the same result could not be as well or even better achieved without it; for MR BEEBE remarks that small-sized males seem sometimes to have the advantage over large ones and these have always a much smaller horn. He remarks also that the female, the cause of the rivalry, invariably escapes during the progress of the contest, so that it results in no advantage to either of the rivals and we are still left without any explanation of the strange outfit. Were these structures really of any importance for offence or defence the extravagances to which they are so liable would assuredly not be found, for grotesqueness is no characteristic of an effective weapon.

The ornament theory offers no solution of the problem. DARWIN'S belief that male beetles, by the possession of horns, are rendered more attractive to the females and those least well furnished with such attractions liable to be rejected by them has been found untenable on account of the poor sight with which the beetles are endowed. This fact must also lead to the rejection of any idea that the structures could serve as recognition marks by which any individual can distinguish a member of its own species from one of another, even if such a supposition were not negatived by the peculiar inconstancy of these features, the dissimilarity of small and large phases and even the occurrence side by side of quite different phases. There is also abundant evidence that female insects are invariably sought for by the males and not *vice versa,* so that recognition marks, if required, would be displayed by the females and not by the males. All the evidence indicates that recognition is by scent and not by sight.

Although some male beetles are certainly enabled to perform special functions better by adaptations of their horns, the fact that females, which, generally unassisted, carry out tasks far more varied and more complicated than any known to be performed by males, are as a rule unprovided with such accessories, does not point to these being in general of any great importance in the beetle economy. Indeed if, as the identical

132

horns of male and female in Passalidae and some other beetles perhaps indicates, horns were originally common to both sexes their usual suppression in the female must be supposed to be due to them constituting a hindrance in the performance of her tasks. Their persistence in males, as well as the frequent absence and sometimes the fantastic exaggeration of the scraper-teeth of their fore-legs and, in many Stag-beetles, the exaggeration of the mandibular teeth, accompanied by the loss of the alternation which formerly gave increased gripping power, may therefore all be regarded as evidences of a usually idle life. Reasons, derived from the frequent absence in males of the marks of wear found in the fore-legs of their females, have been adduced for believing that participation by the male in the tasks of nidification is to be found only where horn-development has not reached a very advanced stage. Grounds for supposing any practical purpose to be served by the horns exist only in a minority of cases.

It may seem remarkable that organs of large size and curious shape should nevertheless be purposeless, but the existence, even in the human body, of useless and even of potentially dangerous structures is well known. If the operation of Natural Selection is unimpeded, such structures, in course of time, if actually harmful, may be eliminated or reduced to harmlessness but, if no effects injurious to the species are entailed, no such result will occur. Many features of this non-useful kind are to be found in so-called correlated characters, that is to say, characters which are bound up with others of greater importance, upon which, in some unexplained way, they depend. It has been shown that horns, not in beetles only but in other animals as well, are associated with large size and that their degree of development depends upon the size of the individual bearing them. Unlike most parts of the body, the size of the horns does not bear a fixed proportion to the size of the body but, when larger and larger examples of the same species are examined, proves to bear a greater and greater proportion to it. In the smallest specimens they are often absent.

The Belgian entomologist LAMEERE has pointed out that a gradation may be observed between the sizes of the horns in different-sized species of beetles, nearly related, similar to that found in different-sized individuals of the same species. We find in various groups, such as the genus *Golofa,* of which several examples are shown upon Plate 10, small species without horns, medium-sized species with medium-sized horns and large-sized species with disproportionately large horns, the different horn-forms pointing to a common origin, but the degree of development determined in every case by the size attained. In a more general way a similar state of things may be observed in whole families, or subfamilies. The Rose-beetles (Cetoniinae), for instance, include large numbers of small kinds, all of which are hornless, as well as many gigantic forms, *Goliathus* (Plate 11, fig. 2), *Mecynorhina* (fig. 4), *Inca* (fig. 1), etc., all of them horned in the male. In the Dynastinae again all the numerous small forms are hornless and the large ones horned, and amongst these are found not only the largest of all existing insects, but the most magnificently horned, like the Atlas and Hercules-beetles. The case is the same in both the Stag-beetles (Lucanidae) and the Longicorns (Cerambycidae), the giants being fantastically horned and the dwarfs hornless. It is true that there are many small horned beetles. In the great genus *Onthophagus,* which contains 1600 different species, most of them horned although few of them are more than half an inch long, the same general rule is found to apply, for the best developed horns are found among the larger members of the genus, smaller horns are borne by the smaller kinds, while others still more tiny have none at all. The tiny species may be actually more ancient than the larger ones or they may have been derived from larger forms by a reversal of the evolutionary process, but the dependence of horn-growth upon body-size is clear. In the course of ages fluctuations in size may have occurred repeatedly but, if so, they have probably always been accompanied by corresponding changes in the horns.

In this close correlation between the size of beetle horns and that of their bearers, both amongst the individuals of a species and the species of a group, we find again a remarkable parallel with higher animals. Just as the largest tusks are borne by the largest elephants and the finest antlers by the most majestic stags, so does the African elephant carry much larger tusks than the smaller Indian elephant, so did the recently extinct Irish deer bear much larger antlers than any of its smaller kin that have survived it. Moreover when we trace back their geological record we find, in

134

elephant and stag, a history of gradually increasing size, accompanied in each case by an increase, though a less gradual one, in the size of tusk and antler, from quite small beginnings to the huge dimensions at last attained. Other great beasts have flourished in past ages whose ancestors, at first of comparatively small size and bearing small horns, have become gradually larger, their horns regularly increasing in relative size. After attaining gigantic bulk, with horns of huge proportions, they have disappeared from the earth and left no representatives. The history of the giant Elephant- and Hercules-beetles has undoubtedly been similar to that of the elephant and stag and their ancestors are represented by the smaller related forms, with much smaller horns, which still exist.

We thus find the cause of exaggerated horn-development in the increasing size of the animal itself. As the generations have succeeded one another a gradual increase of body-size has automatically entailed a less gradual increase in the horns. The explanation of this is therefore to be looked for, not in any advantage to be derived from it, but in the cause of the increase of body-size. It has been noted that amongst the breeds of European cattle long horns are correlated with long hair. It is therefore unnecessary to look for any advantage in the long horns. They are to be explained by the causes that have produced thicker coats. Similarly fantastic horns in beetles can be attributed to a long-continued increase in the size of the insect bearing them. The increase of size may be merely a manifestation of the vigour of the species or it may be explained by the survival of the fittest. Either larger or smaller size, according to circumstances, may be of advantage in the struggle for survival. The relative size of all predacious animals and their prey is always of importance to both. Insects rather too large to be easily dealt with or rather too small to be worth attacking will have an advantage over the rest and a change beginning in either direction may be continued for long periods of time.

In the size-correlation existing between body and horns we can find the explanation of the remarkable compensatory outgrowths which have been described — the emergence of the strange processes beneath the body of the male *Onitis Castelnaui,* Plate 8, fig. 6, *pari passu* with the loss of the head-horns, and the replacement of a large single horn by two small ones in the male *Onthophagus sagittarius,* Plate 5, figs. 13 & 14.

135

The changes undergone by *Pinotus Mormon,* Plate 5, figs. 3 & 4, where the male has more evidently acquired an improved form, are still more readily apprehended.

While increasing size may be an advantage, it may also be a disadvantage. In burrowing insects the diameter of the body determines that of the burrow and even a slight increase entails a large increase in the amount of material to be removed. The labour to be expended in removing it and the need for economy of effort has no doubt been a factor in horn-changes. When the male takes no part in the labour of nidification this factor is absent in his case and, Natural Selection not being brought into operation, the size-factor in the development of the horns is uncontrolled. It is therefore where a great increase in size has occurred and where there is no collaboration between the sexes that we find fantastic extravagance in the male horns.

LAMEERE, speculating upon the origin of the horns in Lamellicorn beetles (L'Evolution des Ornements Sexuels, 1904), and having noticed that a cavity, often with its margins drawn out to form horns, was met with in different groups, formed a conception of an ancient ancestor of the whole sub-order, a contemporary of the giant Dinosaurian reptiles and perhaps in the habit of burrowing into and feeding upon the pithy stems of the Cycads abounding in the Mesozoic age. He supposed both sexes of this ancient insect to have acquired a hollow receptacle upon the back, serving to remove the excavated material from their burrows, and that the margins of this were drawn out at certain points, as seen in the shovel-like ends of the wing-cases of some wood-boring beetles of to-day. He assumed three such projections, one behind and one on each side of the thoracic cavity, while a fourth elevation was situated upon the head. By the exaggeration in the males of these four elevations he considered Lamellicorn horns in all their variety to have originated. All the Lamellicorn beetles without horns he believed to be so, not because they were not acquired but because they have been lost, and Stag-beetles and others with enlarged mandibles have gained these, according to his view, in compensation for their lost horns.

It is true that the ancestry of these beetles can be traced back to the remote age of the Dinosaurs but whether Palaeontology will ever provide confirmation of LAMEERE'S interesting theory is unfortunately very doubtful. His belief that all the different horn-forms now found have originated from four primeval processes cannot be accepted, for they have certainly originated in many different ways. Those upon the head are extensions, sometimes of the front margin, sometimes of the hind margin and sometimes of a ridge formed along the line of junctions of two once separate segments. Those upon the thorax, although often derived from the margins of a cavity, may also be extensions of the outer edges or of the angles. Horny outgrowths, in these and other beetles, are liable to occur in many parts of the body and they may be found in different parts in species rather nearly related as in the great genus *Onthophagus*. It seems likely therefore that their origin is not always to be dated back to the very remote past.

LAMEERE has tried to explain the fact that horns are now mainly confined to male beetles by the abandonment, in that sex, by most of the descendants of the ancient ancestor, of the primitive habit of sharing the labours of the females. Being no longer of use for that purpose the apparatus subsequently became developed for offensive or defensive purposes. But this does not explain the loss of that apparatus by the female nor its retention by many males which have not abandoned the habit of co-operation and I have emphasized the surprising lack of apparent adaptation for combative uses. The occurrence of a hollowed - out thorax, frequent as it is, is almost invariably confined to the males and, if formerly of use in both sexes, it appears strange that it should be retained by these alone. It seems possible that, as LAMEERE suggests, this structure originated as a scoop for clearing the burrow but, as it seems that this has always been the special duty of the male, it is quite likely that it originated in that sex from the flattened area often found in the same region. A beetle found in Celebes, *Ceratoryctoderus Candezei,* Plate 6, fig. 7, with the narrow cylindrical shape of so many wood-borers, has, like the very different *Bolboceras,* a flattened area behind the head and the elevation of its sides has converted this into a hollow, which is enclosed by short horns rising from both head and thorax. In the New Hebrides the nearly related *Enoplus tridens,* fig. 10, which has not the cylindrical form of the burrower, has the thoracic cavity almost obliterated by the approximation of its sides,

137

which, instead of stout elevations, form a pair of rather slender horns, while the raised front of the head is drawn out into a fantastic trident. Are we to conclude that the horns of ancestors of *E. tridens* were useful to them in their burrowing operations, but their descendants having ceased to perform such operations, their horns have lost their useful character and become fantastic; or has the reverse process occurred and horns, at first fantastic, been gradually brought to assume the useful character seen in *C. Candezei?* There is yet another alternative — that incipient horns in males taking part in the burrowing operations have become adapted for use in the task, while in others not so employed, their growth being unrestrained, they have become fantastic.

In an illuminating contribution made to the study of this problem by Prof. C. CHAMPY (Charactères Sexuels et Hormones, 1924) he has called attention to the similarity in the characteristics of the external male features in animals so diverse as beetles and butterflies, crabs, fishes, frogs and newts, lizards, birds and mammals. His researches on the subject of the influence of the thyroid gland secretion upon the growth of vertebrate animals convinced him that an internal secretion or hormone also controls the development of all these sex-features, including not only measurable bodily structures like horns but such intangible characteristics as the more vivid colours of male birds and butterflies and the songs of male birds, crickets and grasshoppers. CHAMPY'S theory, although supported by weighty arguments, was a bold one, for internal secretions of the kind at that time were known to exist only in vertebrate animals; but recent discoveries by WIGGLESWORTH and others that various functions in insects are actually controlled by secretions discharged into the blood-stream have removed the only serious objection to its acceptance.

As horns are only one form of the outgrowths that may appear in almost any part of the insect-body, so outgrowths in general are only one kind of manifestation of all the varied phenomena known as second-ary sexual characters. Why these take sometimes one form and sometimes another is a problem upon which as yet hardly any light has been shed. CHAMPY has pointed out (*op. cit.,* p. 233) that it is the unessential parts of the body that are affected. The enormous enlargement of the male mandibles does not occur in the carnivorous beetles, which need them

138

for seizing their prey, but in stag-beetles, whose food consists of vegetable juices. Similarly the canine teeth of the Carnivora, which also are of special importance to them, have not become tusks; this has happened in male animals such as swine, some deer, etc., to which these teeth were of no special importance, while the incisor teeth that have become great tusks in the elephant and the narwhal, not being required for cropping herbage, had probably ceased to serve any particular purpose. It seems likely that in the stag-beetles the mandibles of the females have always been of use and only those of the males were without employment. In many horned beetles also, where the front edge of the head of the male has become produced into a horn, this part was of use to the female in burrowing but served no purpose in the other sex.

Thus not use but disuse appears to have determined the first inception of such structures and any use they may now have has been acquired later. The assumption underlying so much that has been written on the subject, that some use must be assumed from the fact of their existence, is quite unjustified. As features correlated with size, their is no more reason for assuming a use than there is for assuming usefulness for the deafness usually accompanying whiteness of fur and blue eyes in male cats mentioned by DARWIN (Origin of Species, 1902 ed., p. 13). There is an intimate but unexplained connection between these apparently unrelated qualities. Increasing size in the insect has at some point entailed the appearance of horns, which have the property of continuing to develop, but at a faster rate, so long as the species continues to grow larger, and only if they chance to become in some way useful can their manner of development be affected by the adaptation produced by Natural Selection. As this close correlation between horn-development and the size of the animal is found to exist in insect and mammal alike and, as size in mammals has been found to be controlled by a glandular secretion carried in the blood, it is reasonable to conclude that somewhat similar secretions, or hormones, are involved in both cases.

THOMSON & GEDDES (The Evolution of Sex, 1889, p. 24) have suggested a resemblance to the getting rid of waste products in the growth of the stag's antlers and LAMEERE (L'Evolution des Ornements Sexuels, 1904) has attempted to explain the horns of male beetles as the equivalent of

139

the material expended by the female in the production of eggs. But the phenomena it is desired to explain are immensely varied and by no means confined to bodily outgrowths. It has also been pointed out by THOMSON & GEDDES that, as greater passivity, often accompanied by comparative longevity, characterizes the female sex throughout Nature, so a more concentrated vitality and expenditure of energy are distinctive of the male, and this is manifested in all the vast range of secondary sexual characters, outgrowths of all kinds, as well as brilliant colours, songs, scents, etc. This was recognized by WALLACE in his contention that the brilliant colours and showy plumage of so many male birds are the outcome of their superabundant vitality.

WALLACE also pointed out how Natural Selection has emphasized the difference between the sexes by often repressing the tendency in the female to acquire the bright plumage of the male and bringing about instead a protective coloration. The conspicuousness resulting from the brighter plumage of the male would obviously be specially disadvantageous to the hen-bird when sitting upon her nest (eggs or young) and thus is explained the frequent assimilation of her plumage to the environment, as in that of the hen-pheasant and so many other birds. The same phenomenon is very apparent in many beetles, such as the Hercules-beetle, of which the female has a rough surface of dull brown, harmonizing with the ground in which she must burrow to deposit her eggs, while the male is very glossy, his immense horns and anterior parts brilliant black and the hinder half shining green. Among the branches of a tree he may not be very easily discovered, but upon the ground, unlike the female, he would be conspicuous. One of the most lovely of all beetles, the glistening sky-blue and silver *Hoplia coerulea,* found in southern France, has a female of an earthen brown colour. The significance of such striking sex-differences, which are very numerous, is shown in the remarkable facts noted by Wallace that, in certain instances where the female bird has a brighter plumage than the male, such as the Dotterel and the Phalaropes, the habits of the two sexes are actually reversed, the male incubating the eggs and the female being the more pugnacious bird; while in those groups in which the male and female are both brightly coloured, the Parrots, Woodpeckers, Kingfishers and others, the nest is

140

invariably placed in a hole or in such a situation that the colours of the sitting bird are invisible.

It may be expected that the rare cases in which male and female beetles have both highly developed horns will be explained when their habits are sufficiently known. The great Coprid, *Phanaeus lancifer,* whose manner of life is quite unknown, is a rare example of the kind amongst beetles and the Reindeer may be regarded as such an exception amongst the mammals, to some extent accounted for by its existence in an arctic climate beyond the tree-line and beyond the range of large carnivorous animals.

The infrequent occurrence of well-developed horns in beetles of both sexes, although the male characters are capable of transference to the other sex, shows that there is an influence stronger than that of heredity which in the female has usually restrained it; and those cases in which the horns of the male have become adapted to serve practical ends indicate that this influence is that of Natural Selection, by which the more efficient individuals survive to become the parents of the next generation. Considering the exceptional variability of these structures it can scarcely be doubted that, where they become of any consequence in relation to the habits, any variations in the direction of greater serviceability will have the best chance of survival and very extravagant forms will persist only if no adverse effect upon the well-being of the race is involved. WALLACE has called attention to the exceptional abundance of such extravagantly decorated male birds as the Peacock and Birds of Paradise; and some of the giant horned beetles seem to be equally remarkable for their abundance in the male sex. The swarming of the males of *Dynastes Tityus* has already been referred to and, although the more retiring habits of the female may be partly responsible for their apparently smaller numbers, there seems little doubt that in many, and perhaps in most, of the extravagantly horned beetles males are actually much in excess of females. By tethering a female of the common Stag-beetle, *Lucanus cervus,* to a tree-stump 75 males were attracted to the spot and collected in a hour and a half (CORNELIUS, *Stett. Ent. Z.,* 1868, p. 24). Where there is a great excess of males a considerable mortality amongst them will be without the effect upon the next generation that it would have if it occurred

141

amongst the females. Such a disproportion between the two sexes has not been observed in any of the beetles known to work in couples.

We have found then that horns in beetles, as in other animals, and like similar features such as the exaggerated plumes, beaks, casques, wattles, etc., in birds, are especially characteristic of the males of the largest forms in any group; that they have the property of increasing with increasing size in the animal but at a more rapid rate; that they are found to affect parts of the body the precise aspect of which is of no great importance; and, their development being a consequence of the increasing body-size, it is unnecesssary to seek for any special function, in their beginnings at least. If, in the course of later development, any advantage or disadvantage to the species should result from their existence, elimination of less advantageous forms in the struggle for survival will effect changes in a more advantageous direction. If, on the other hand, no advantage or disadvantage to the species is entailed and increase in body-size continues, the disproportionate increase of the outgrowth must ultimately produce a structure of fantastic proportions relatively to the size of the animal. The process must have a definite limit. The great proportionate size of the horns in the largest beetles now living makes it improbable that any of much larger size and, as a consequence, with very much larger horns, can ever have existed. In some of the horned giants flight must already have become difficult and the continuation of the process would result in all locomotion becoming impeded and the perpetuation of the species at last endangered. Although it may seem strange that appendages should continue to develop in spite of increasing inconvenience to the individual burdened with them, we have to admit that this has happened visibly in elephants, stags and giant sheep but, owing to the fact that such inconvenience is spared the females and young, it does not jeopardize the species. Unless the survival of the young is endangered by such inconveniences they are without effect in the struggle for existence.

If structures, which for the individual although not for his species are actual encumbrances, continue to persist it is not remarkable that those that are merely useless do not disappear. Plenty of instances can be found of quite useless structures in the body which have existed throughout vast periods of time. The bones representing the hind limbs in the skeleton of a whale and those in the human skeleton representing the tail are familiar instances.

It is easily seen that horns have arisen from pre-existing prominences or ridges in different parts of the body. The extension of these and the increasing body-size of which it is the consequence can only occur if body-building material is in abundance and horn development implies a surplus of the horny substance of which the exterior of an insect consists. Although small outgrowths may be found in many parts of the body, there are only a few regions in which greater outgrowths are possible without interference with vital functions. In mammals only such teeth as are not employed in getting or masticating food become tusks, in male birds the extravagantly developed plumes of peacock or pheasant are not feathers necessary for flight but comparatively unimportant ones; in male beetles the strong mandibles, without employment for practical purposes, the hard scraper-tibiae of the fore-legs, not required, like those of the female, for digging, are often affected. But it is upon the upper surface of the head and thorax that outgrowths can develop most freely without serious interference with vital functions. When two horny plates, formerly separated, have become united by the deposition of horny substance between them, as has happened to the dorsal and lateral plates in the prothorax of a beetle or those the union of which can be traced at the top of the head, the deposition seems often to continue until a ridge is formed where there was formerly a groove. Perhaps because a process long continued is not easily discontinued, such a ridge may become very much accentuated and, contracting as it grows outward, tapering to a point, form a horn; or the production of the ridge at its two ends may result in the formation of a pair of horns. If the process continues the horns may give rise to secondary outgrowths, become forked, toothed or branched in many different ways. If it should happen that they come to have any possible relation to the propagation of the species the operation of Natural Selection will eliminate the least advantageous forms and bring about gradual improvement; if they are without any such relation Natural Selection will not be brought into operation. In female insects the utmost efficiency in every respect is necessary for perpetuation of the species, ability to evade capture and survive long enough to accomplish all the operations required to ensure the survival of offspring, to find the proper environment for the eggs, place them in safety and make provision for the needs of the future generation. In the male all this is in most insects unnecessary. His sole function is to fertilize the eggs. The number of males is often far greater than is required for the purpose and a drastic reduction may have no ill effects whatever. Ability to escape enemies is of little importance, a very brief survival is sufficient. He has inherited organs of which he has no real need and whether they are efficient or not is unimportant. Ability to locate the female and ardour to pursue her are the qualities needed for the survival of the race and others with no bearing upon that supreme object are not affected by Natural Selection. The horns of a beetle, the size of which is increasing gradually as generations succeed one another, will as a result become more and more disproportionate in size, regardless of the fact that they may be quite useless, and the absence of the restraining and moulding influence of Natural Selection will be a contributory cause of the acquisition of fantastic forms. It is the influence of Natural Selection which, by insuring the fitness of every part of the body for its particular purpose, maintains the general appearance of congruity, whereas the absence, whether it is actual or only apparent to us, of such fitness produces upon our minds the effect of incongruity. It is the tendency possessed by horns and many other features of a similar kind to appear in those forms in any group which are increasing in size, together with the continually increasing disproportion which accompanies that process, which have resulted in the giants in many groups of animals, beetles as well as others, presenting so many of the most notable examples of the fantastic to be found in Nature.

AUTHORS INDEX

145

GENERAL INDEX

Ambrosia-beetles 12 et seq., 36 et seq., 71, 119.
Antelopes 109, 113.
Anthribid beetles 18.
Ants 5, 68, 119, 123, 126.
Atlas-beetle 17, 18, 60, 101, 134.
Babyrusa 4.
Bees 5, 68, 119, 120.
Beetles,
  antennal modifications 15, 18.
  ball-rolling 27, 129.
  caudal appendages 12.
  characters of order 4, 5, 6.
  compensatory appendages 81, 128.
  egg-laying 20.
  food 6, 13, 20.
  habits 5, 24 et seq.
  larval stridulation 30.
  legs, modifications 14, 15, 19, 28, 84, 143.
  legs, teeth wearing 52 et seq.
  nest-building 92.
  numbers 3.
  predaceous 7, 20.
  sensory organs of females 24.
  sensory organs of males 24.
  subdivisions of order 7.
  vegetarian 8, 20.
  wing-cover modifications 15, 18, 82.
Birds, 2, 113, 115, 124, 125, 126, 140, 141, 142.
  of Paradise 141.
  plumage 2, 143.
Black Coconut-beetle 58.
Bower-birds 118.
British Museum, collections 53.
Bugs 5.
Butterflies 5, 119, 120, 124, 138.
Carnivora 139.
Cattle 3, 135.
Cats, deafness 139.
Chafers 8, 25.
Chameleons 113.
Chamois 109.
Cockroaches 5.

Coloration, protective 17.
Cooperation of sexes 68 et seq.
Crabs 116, 138.
Crickets 138.
Crustaceans 3.
Deer, 114, 139.
  Red, 111.
Dinosaur 136.
Dogs, size 83.
Dor-beetles 8, 30, 40 et seq.
Dotterel 140.
Dragonflies 120.
Elephant 4, 109 et seq., 116, 134, 139, 142.
  African, 112, 134.
  Indian, 112, 134.
Elephant-beetles 21, 59, 61, 135.
Elephant dung-beetle 111.
Elk, 114.
  Irish, 111, 114, 134.
Eyes, beetles 120 et seq.
  butterflies & moths 119.
  dragonflies 120.
  house-fly 120.
  insects 118 et seq.
Fireflies 8.
Fishes 2, 3, 113, 116, 138.
Flies 5, 120.
Frogs 138.
Fungi, Ambrosia 13.
Gemsbuck 113.
Geological evidence 111.
Giraffes, fossil 112.
Glowworms 8.
Goliath-beetles 8, 63, 95, 105, 109.
Grasshoppers 5, 138.
Ground-beetles 7, 10.
Harlequin-beetle 19, 82.
Hercules-beetles 18, 21, 59, 96, 101, 109, 112, 121, 134, 135, 140.
Hormones 138 et seq.
Hornbill 2.
Horns, 2 et seq., 17, 104, 109, 138, 142.
  beetles, 2, 4, 6 et seq., 9 et seq., 66, 104, 116, 131 et seq., 139, 141

146

Clytrinae, Chrysomelidae. 9.
Coelosis Hope, Dynastinae. 56.
Colophon Gray, Lucanidae. 86.
Coprinae = Scarabaeinae, Scarabaeidae.
Copris Geoffroy, Scarabaeinae.
   33, 51, 70, 71, 74, 75, 101, 107, 119.
C. draco Arrow. 73.
C. hispanus (Linnaeus). 50, 72.
C. lunaris (Linnaeus). 49 et seq., 53,
   72, 73. Pl. 5, f. 5, 6.
Crossotarsus Chapuis, Platypodidae. 13.
Cucujidae, Coleoptera. 85.
Curculionidae, Coleoptera. 9.
Cyphonocephalus Westwood, Cetoniinae.
   94.
C. smaragdulus Westwood = *olivaceus*
   (Dupont). Pl. 12, f. 4, 5.
Cytorea Castelnau, Erotylidae, Coleop-
   tera. 14.
Deinophloeus ducalis Sharp, Cucujidae.
   85.
Dicaulocephalus Gestro, Rutelinae.
D. Fruhstorferi Felsche. 85.
Diceros malayanus (Wallich), Cetoniinae.
   Pl. 12, f. 9.
Dicranocephalus Burmeister, Cetoniinae.
   95.
D. Bowringi Pascoe. Pl. 12, f. 1, 2.
Dicranorhina Gralli (Buquet) (*Eudicella*
   White), Cetoniinae. Pl. 12, f. 10.
Didrepanephorus bifalcifer Wood-Mason,
   Rutelinae. 84. Pl. 14, f. 10.
Diloboderus Abderus (Sturm), Dynasti-
   nae. 55, 111.
Dipelicus Cantori Hope, Dynastinae.
   21.
D. Geryon (Drury). 21, 96. Pl. 13, f.
   4.
Diurus forcipatus (Westwood), Brenthi-
   dae. 16, 18. Pl. 4, f. 5, 6.
Dorcus Macleay, Lucanidae. 77.
D. arfakianus (Lansberge) (*Eurytrache-
   lus* Thomson, J.). 87. Pl. 14, f. 1—
   7.
D. Arrowi (Gravely) (*Cladognathus* Bur-
   meister). 90.

D. Boileaui (Didier) *(Rhaetulus)*. Pl. 9,
   f. 3.
D. elegans Parry (*Hemisodorcus* Thom-
   son, J.). Pl. 9, f. 8.
D. forceps (Vollenhoven) (*Prosopocoe-
   lus* Westwood). 89. Pl. 15, f. 20—
   22.
D. giraffa (Fabricius) (*Cladognathus*).
   90. Pl. 15, f. 10, 11.
D. Mellyi see *Homoderus.*
D. polymorphus Arrow (*Prosopocoelus
   Parryi* Boileau). 89. Pl. 15, f. 23—
   25.
D. Reichei Hope *(Eurytrachelus)*. 88.
   Pl. 15, f. 12, 13.
Dorysthenes Vigors, Cerambycidae. 85.
D. rostratus (Fabricius). 16.
Drepanocerus Kirby, W., Scarabaeinae.
D. Kirbyi Kirby. 98.
D. setosus (Wiedemann). 98.
D. sinicus Harold. 98.
Dynastes Kirby, W., Dynastinae.
   59, 82.
D. dichotomus (Linnaeus) *(Allomyrina*
   Arrow). 104. Pl. 13, f. 7.
D. Gideon (Linnaeus) (*Xylotrupes*
   Hope). 60, 97, 101, 102, 104, 106.
   Pl. 8, f. 3, 4.
D. Granti Horn. 63, 103.
D. Hercules (Linnaeus). 96, 97, 101,
   102, 103, 104, 106. Pl. 2.
D. Hyllus Chevrolat. 103.
D. inarmatus Sternberg. 60, 104.
D. Neptunus (Quensel). 96, 97, 103,
   104. Pl. 1a.
D. Tityus (Linnaeus). 59, 103, 141.
Dynastinae, Scarabaeidae. 8, 54 et seq.,
   71, 78, 92, 98, 121, 127.
Dysantes elongatus (Redtenbacher),
   Tenebrionidae. Pl. 4, f. 15.
Dytiscidae, Coleoptera. 8.
Enema Hope, Dynastinae. 56.
E. Pan (Fabricius). 56 et seq., 77, 105.
Enoplus tridens Montrouzier (*Hoploryc-
   toderus* Prell), Dynastinae. 137. Pl.
   6, f. 10.
Erotylidae, Coleoptera. 92.

149

Euchirus longimanus (Linnaeus), Euchirinae. 15, 19.
Eupatorus Beccarii Gestro, Dynastinae. Pl. 13, f. 8.
Eurysternus Dalman, Scarabaeinae. 98, 99.
Exorides equicaudatus Marshall, Curculionidae. 16.
Figulus MacLeay, Lucanidae. 32, 86.
Fruhstorferia sexmaculata Kraatz, Rutelinae. 84. Pl. 8, f. 7—9.
Galerita Fabricius, Carabidae. 10.
Geotrupes Latreille, Geotrupinae. 8, 30, 44, 45, 67, 119.
G. Momus (Olivier) (*Typhaeus* Leach). 45. Pl. 7, f. 3, 4.
G. Sharpi Jordan & Rothschild (*Enoplotrupes* Lucas, H.). 45.
G. Typhoeus (Linnaeus) (*Typhaeus*). 40 et seq., 101. Pl. 7, f. 1, 2.
Geotrupinae, Scarabaeidae. 8, 40, 49, 55, 70, 121.
Goliathus Lamarck, Cetoniinae. 134.
G. albosignatus Boheman. 64, 95, 105. Pl. 11, f. 3.
G. Druryi MacLeay (*G. Goliathus* Drury). 63 et seq.
G. Fornasinii (Bertolini) (*Fornasinius* Bertolini). 95, 105. Pl. 11, f. 2.
Golofa Hope, Dynastinae. 59, 60, 96, 103, 131, 134.
G. argentina Arrow. 96, 103. Pl. 10, f. 6.
G. claviger (Linnaeus). 96, 97, 103. Pl. 10, f. 2—4.
G. Eacus Burmeister. 103.
G. Guildingi Hope. 103. Pl. 10, f. 5.
G. imperialis Thomson, J. 103.
G. inermis Thomson, J. 103. Pl. 10, f. 7.
G. Porteri Hope. 21, 97, 103. Pl. 10, f. 1.
Gymnopleurus Illiger, Scarabaeinae. 75.
G. malleolus Kolbe. 75.
G. thoracicus Harold. 75.
G. virens Erichson. 75.

Heliocopris Hope, Scarabaeinae. 51, 107.
H. dominus Bates. 107, 128.
H. gigas (Olivier). 21, 107, 111. Pl. 7, f. 15, 16.
Heterogomphus Burmeister, Dynastinae. 56.
Heteromera, Coleoptera. 8.
Hexarthrius Parryi Hope, Lucanidae. 33.
Histeridae, Coleoptera. 8.
Homoderus Mellyi Parry, Lucanidae. 87. Pl. 14, f. 11—13.
Hoplia coerulea (Drury) Melolonthinae. 17, 140.
Hydrophilidae, Coleoptera. 8.
Hymenopus Audinet-Serville, Orthoptera. 119.
Inca Lepeletier & Serville, Scarabaeidae: Trichiinae. 134. Pl. 11, f. 1.
Laemophloeus Castelnau, Cucujidae. 85.
Lamellicornia, Coleoptera. 8, 93.
Lamprima Latreille, Lucanidae. 90.
L. Adolphinae (Gestro) (*Neolamprima* Gestro). 90, 91. Pl. 9, f. 4—7.
L. Latreillei MacLeay. 89, 90.
Lampyridae, Coleoptera. 8.
Lethrus Scopoli, Geotrupinae. 20, 39.
L. apterus (Laxmann). 39 et seq., 67.
Liatongus Reitter, Scarabaeinae. 98.
L. mergaceras (Hope). 98.
L. monstrosus (Bates) 77. Pl. 6, f. 1, 2.
L. Rhadamistus (Fabricius). 98.
L. vertagus (Fabricius). 76.
Lucanidae, Coleoptera. 8, 20, 25, 26, 32 et seq., 86, 93, 122, 127, 134.
Lucanus Scopoli, Lucanidae. 33.
L. cervus (Linnaeus). 20, 33, 34, 86 et seq., 121, 141. Pl. 8, f. 1, 2.
L. Fortunei Saunders. Pl. 14, f. 8, 9.
Lycomedes ramosus Arrow, Dynastinae. 104.
L. Reichei Brême. 104.
Macrodontia Audinet-Serville, Cerambycidae. 20, 121.
M. cervicornis (Linnaeus). 6, 82 et seq. Pl. 3.

BIBLIOGRAPHY.

ARROW, G. J. 1899. "On Sexual Dimorphism in Beetles of the family Rutelidae". *Trans. ent. Soc.*, London, **1899**, p. 255—269.

— 1910—50. Fauna of British India, Coleoptera Lamellicornia, London, 4 vols.

— 1943. "On the Genera and Nomenclature of the Lucanid Coleoptera, and descriptions of a few new species". *Proc. R. ent. Soc.*, London, (A) **15**, p. 133—143.

— 1944. "Polymorphism in giant beetles". *Proc. zool. Soc.*, London, (A) **113**, (1943) p. 113—116, 1pl.

BATES, H. W. 1863. "The Naturalist on the River Amazons". London. Several editions.

BATESON, W. & BRINDLEY, H. H. 1892. "On Some cases of Variation in Secondary Sexual Characters, statistically examined". *Proc. zool. Soc.* London, **1892**, p. 585—594.

BEEBE, W. 1944. "The function of secondary sexual characters in two species of Dynastidae (Coleoptera)". *Zoologica,* New York, **29**, p. 53—8, 5 pls.

BEESON, C. F. C. 1917. "The life-history of *Diapus furtivus* Sampson". *Ind. Forest Rec.* **6**, p. 1—29, 2 pls.

BENNETT, A. L. 1899. "Notes on the habits of *Goliathus druryi*". *Proc. ent. Soc.*, London, **1899**, p. xi—xiii.

BRISTOWE, W. S. & LOCKET, G. H. 1926. "The courtship of British Lycosid Spiders, and its probable significance". *Proc. zool. Soc.*, London, **1926**, p. 317—347, 10 ff.

CHAMPY, C. 1924. "Sexualité et hormones". Paris.

CHAPMAN, T. A. 1868. "Note on the habits of *Sinodendron cylindricum* during oviposition etc." *Ent. mon. Mag.* **5**, p. 139—141.

CORBIN, G. B. 1874. "The Dor-beetle at work". *Entomologist*, **7**, p. 132—135.

CUNNINGHAM, J. T. 1900. "Sexual Dimorphism in the Animal Kingdom". London.

DAGUERRE, J. B. 1931. "Costumbres nupciales del *Diloboderus abderus* Sturm". *Rev. Soc. ent. argent.,* Buenos Aires, **3**, p. 253—256, 2 ff.

DARWIN, C. 1839. "Journal of Researches". London. Many editions.

— 1871. "The Descent of Man". London. Many editions.

DOANE, R. W. 1913. "The rhinoceros beetle (*Oryctes rhinoceros* L.) in Samoa". *J. Econ. ent.,* Concord, N.H. **6**, p. 437—442, 2 pls.

FABRE, J. H. 1919—23. "Souvenirs Entomologiques". Paris, 11 vols. (Many editions; English translations of many parts by A. T. de Mattos).

HEYMONS, R. 1929. "Ueber die Biologie der *Passalus* - Käfer". *Z. Morph. Oekol. Tiere,* Berlin, **16**, p. 74—100, 14 ff.

HINGSTON, R. W. G. 1933. "The Meaning of Animal Colour and Adornment". London.

HUBBARD, H. G. 1896. "Ambrosia beetles". *Yearb. U.S. Dept. Agric.,* Washington, **1896**, p. 421—431.

HUXLEY, J. 1932. "Problems of Relative Growth". London.

— 1938. "Darwin's theory of sexual selection and the data subsumed in it, in the light of recent research". *Amer. Nat.,* **72**, p. 416—433.

JUDULIEN, F. 1899. "Quelques notes sur plusiers Coprophages de Buenos Aires". *Revist. Mus. La Plata*, **9**, p. 371—380.

LAMEERE, A. 1904. "L'Evolution des ornaments sexuels". *Bull. Ac. Belgique*, **1904**, p. 1327—1364.

MAIN, H. 1917. "On rearing beetles of the genus *Geotrupes*". *Proc. S. Lond. ent. Soc.*, **1916-7**, p. 18—22, pl. 1.

MANEE, A. H. 1908. "Some observations at Southern Pines, N. Carolina". *Ent. News*, Philadelphia, **19**, p. 286—9.

— 1908. "Some observations at Southern Pines, N. Carolina : Three mound builders." *loc. cit.* **19**, p. 459—62, 2 pls.

MORLEY, B. D. W. 1940. "An artificially produced multiple mixed colony of Ants (Hym.)". *Proc. R. ent. Soc.*, London, (A) **15**, p. 103—4.

MULSANT, E. 1842. "Histoire naturelle des Coléoptères de France, Lamellicornes". Lyon.

OHAUS, F. 1900. "Bericht über eine entomologische Reise nach Centralbrasilien". *Stettin ent. Z.*, **61**, p. 204—45.

— 1908. "Bericht über eine entomologische Studienreise in Südamerika". *loc. cit.*, **70**, 1909, p. 1—139.

PEARSE, A. S. & others. 1936. "The Ecology of *Passalus cornutus* Fabricius, a beetle which lives in rotting logs." *Ecol. Monogr.*, Durham, N.C., **6**, p. 455—90, 43 ff.

PECKHAM, G. W. & E. G., 1890. "Additional observations on Sexual Selection in Spiders of the Family Attidae, with some Remarks on Mr. Wallace's Theory of Sexual Ornamentation". *Nat. Hist. Soc.*, *Wisconsin*, **1890**, p. 117—51.

REICHENAU, W. VON. 1881. "Ueber den Ursprung der secundären männlichen Geschlechtscharaktere, insbesondere bei den Blatthornkäfern." *Kosmos*, **10**, p. 172—94, pl. v.

SAUNDERS, C. J. 1936. "A note on *Odontaeus armiger* Scop". *Ent. mon. Mag.* **72**, p. 178.

SCHREINER, J. 1906. "Die Lebensweise und Metamorphose des Rebenschneiders oder grossköpfigen Zwiebelhornkäfers (*Lethrus apterus* Laxm.)" *Horae Soc. ent. Rossicae*, **37**, p. 197—208, 1 pl.

STROHMEYER, H. 1911. "Die biologische Bedeutung sekundärer Geschlechtscharaktere am Kopfe weiblichen Platypodiden." *Ent. Blätter*, Berlin, **7**, p. 103—7.

THOMSON, J. A. & GEDDES, P. 1889. "The Evolution of Sex". London.

WALLACE, A. R. 1868. "The Malay Archipelago". London, several editions.

— 1870. "Natural Selection". London. Several editions.

— 1878. "Tropical Nature". London. Several editions.

— 1889. "Darwinism". London.

WHEELER, W. M. 1923. "Social Life among the Insects". London.

Plates

All the figures are natural size unless otherwise stated.

PLATE 1.

Chalcosoma Atlas, male and female.

Dynastes Neptunus, male and female.

PLATE 2.

Dynastes Hercules, male.

PLATE 3.

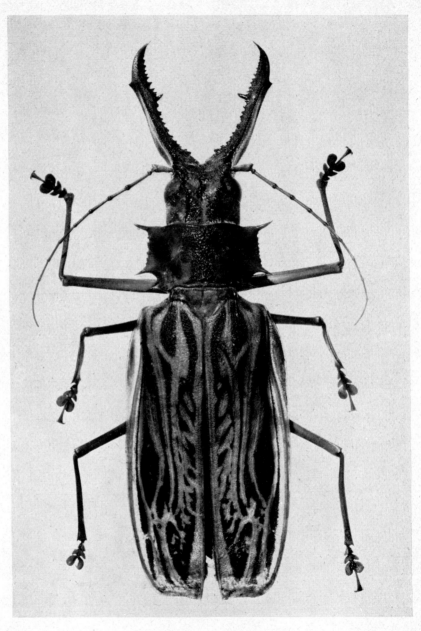

Macrodontia cervicornis, male.

PLATE 4.

Fig. 1. *Cephalobarus macrocephalus,* male.
    2.   ditto,   female.
    3. *Brenthus armiger,* male.
    4.   ditto,   female.
    5. *Diurus forcipatus,* male.
    6.   ditto,   female.
    7. *Mecopus spinicollis,* male.
    8. *Valgus hemipterus,* male.
    9.   ditto,   female.
  10. *Prophthalmus potens,* male.
  11.   ditto,   female.
  12. *Mecomastyx Montraveli,* male.
  13.   ditto,   female.
  14. *Atasthalus spectrum,* male.
  15. *Dysantes elongatus.*
  16. *Bolitoxenus Taprobanae,* male.
  17. *Mechanetes cornutus.*

Of figs. 15 and 17 both sexes are alike.
Figs. 14 to 17 are slightly enlarged.

PLATE 4.

PLATE 5.

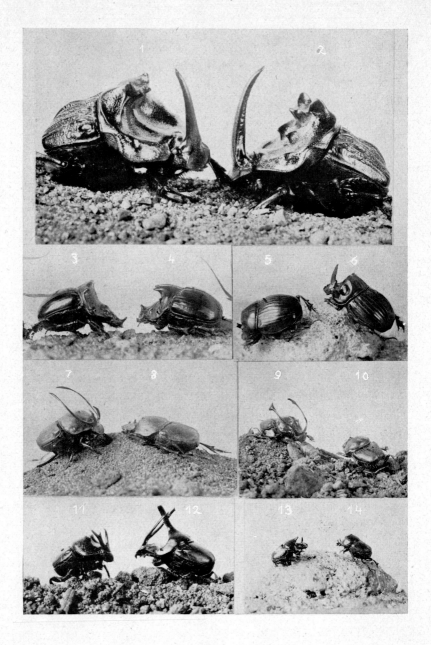

PLATE 5.

Fig. 1. *Phanaeus lancifer,* female.
    2.   ditto, male.
    3. *Pinotus Mormon,* male.
    4.   ditto, female.
    5. *Copris lunaris,* female.
    6.   ditto, male.
    7. *Onthophagus gibbiramus,* male.
    8.   ditto, female.
    9. *O. rangifer,* male.
  10.   ditto, female.
  11. *O. imperator,* female.
  12.   ditto, male.
  13. *O. sagittarius,* female.
  14.   ditto, male.

PLATE 6.

Fig. 1. *Liatongus monstrosus,* male.
    2.   ditto,   female.
    3. *Pseudoryctes dispar,* male.
    4.   ditto,   female.
    5. *Strategus Simson,* male.
    6.   ditto,   female.
    7. *Ceratoryctoderus Candezei,* male.
    8. *Trichogomphus robustus,* male.
    9.   ditto,   female.
   10. *Hoploryctoderus tridens,* male.

PLATE 6.

PLATE 7.

PLATE 8.

Fig. 1. *Lucanus cervus,* male.
    2.   ditto,   female.
    3. *Dynastes Gideon,* male.
    4.   ditto,   female.
    5. *Onitis Castelnaui,* female.
    6.   ditto,   male.
    7. *Fruhstorferia 6-maculata,* male.
    8.   ditto,   female.
    9.   ditto,   male.

PLATE 8.

PLATE 9.

PLATE 9.

Fig. 1. *Chiasognathus Granti,* male.
  2.   ditto, female.
  3. *Dorcus Boileaui,* male.
  4. *Lamprima Adolphinae,* male.
  5.   ditto.
  6.   ditto, female.
  7.   ditto, male.
  8. *Dorcus elegans,* male.

PLATE 10.

Fig. 1. *Golofa Porteri,* male.
2. *G. claviger,* female.
3. ditto, male.
4. ditto.
5. *G. Guildingi,* male.
6. *G. argentina,* male.
7. *G. inermis,* male.

PLATE 10.

PLATE 11.

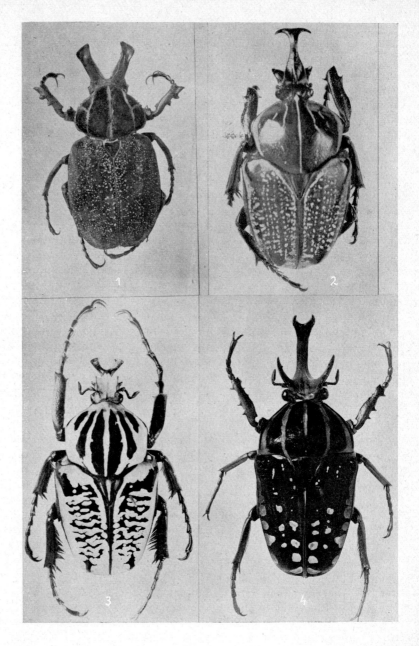

PLATE 11.

Fig. 1. *Inca* sp.

2. *Goliathus Fornasinii,* male.
3. *G. albosignatus,* male.
4. *Mecynorhina Polyphemus,* male.

PLATE 12.

PLATE 13.

PLATE 13.

Fig. 1. *Catharsius molossus,* male.
    2.   ditto, female.
    3. *Pinotus nutans,* male.
    4. *Dipelicus Geryon,* male.
    5. *Catharsius Bradshawi,* male.
    6. *C. platypus,* male.
    7. *Allomyrina dichotoma,* male.
    8. *Eupatorus Beccarii,* male.

PLATE 14.

Figs. 1—6. *Dorcus arfakianus,* males.
Fig. 7.   ditto, female.
     8. *Lucanus Fortunei,* male.
     9.   ditto, female.
     10. *Didrepanephorus bifalcifer,* male.
     11. *Dorcus Mellyi,* female.
     12.   ditto, male.
     13.   ditto.

PLATE 14.

PLATE 15.

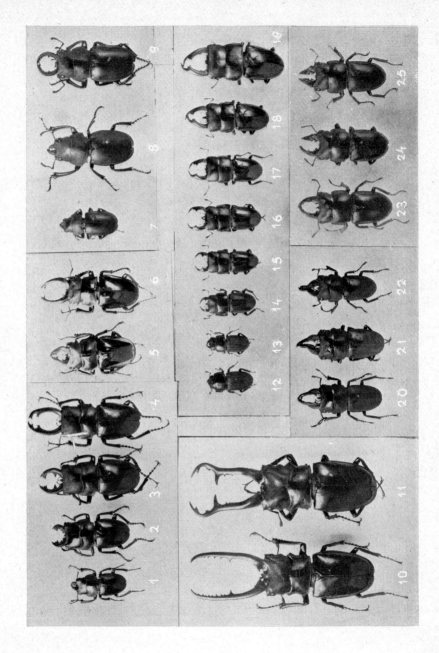